T0146368

Modeling, Simulation, and Operations Analysis in Afghanistan and Iraq

Operational Vignettes, Lessons Learned, and a Survey of Selected Efforts

Ben Connable, Walter L. Perry, Abby Doll, Natasha Lander, Dan Madden

Prepared for the Office of the Secretary of Defense, Cost Assessment and Program Evaluation

This research was sponsored by OSD-CAPE and conducted within the International Security and Defense Policy Center of the RAND National Defense Research Institute, a federally funded research and development center sponsored by the Office of the Secretary of Defense, the Joint Staff, the Unified Combatant Commands, the Navy, the Marine Corps, the defense agencies, and the defense Intelligence Community under contract W74V8H-06-C-0002.

Library of Congress Cataloging-in-Publication Data

ISBN: 978-0-8330-8211-4

The RAND Corporation is a nonprofit institution that helps improve policy and decisionmaking through research and analysis. RAND's publications do not necessarily reflect the opinions of its research clients and sponsors.

Support RAND—make a tax-deductible charitable contribution at www.rand.org/giving/contribute.html

RAND® is a registered trademark.

Photo by Cpl. Ruben D. Maestre

© Copyright 2014 RAND Corporation

RAND OFFICES
SANTA MONICA, CA • WASHINGTON, DC
PITTSBURGH, PA • NEW ORLEANS, LA • JACKSON, MS • BOSTON, MA
DOHA, QA • CAMBRIDGE, UK • BRUSSELS, BE
www.rand.org

Preface

The Office of the Secretary of Defense, Cost Assessment and Program Evaluation (OSD-CAPE) asked RAND to conduct a lessons learned examination of analysis, modeling, and simulation in Operation Enduring Freedom (OEF), Operation Iraqi Freedom (OIF), and Operation New Dawn (OND) to help improve resource allocations in these activities. This effort will identify decisions at all levels of counterinsurgency and, more broadly, irregular warfare that could benefit from more extensive and rigorous modeling and simulation. It also identifies ways in which analysts have attempted to address these decisions, describes many of the models and tools they employed, provides insight into the challenges they faced, and suggests ways in which the application of modeling, simulation, and analysis might be improved for current and future operations. Interviews and review of analytic activities focused on the experience of analysts and their consumers in U.S. forces in OEF, OIF, and OND from late 2001 through early 2012.

This research was sponsored by OSD-CAPE and conducted within the International Security and Defense Policy Center of the RAND National Defense Research Institute, a federally funded research and development center sponsored by the Office of the Secretary of Defense, the Joint Staff, the Unified Combatant Commands, the Navy, the Marine Corps, the defense agencies, and the defense Intelligence Community.

For more information on the International Security and Defense Policy Center, see http://www.rand.org/nsrd/ndri/centers/isdp.html or contact the Director (contact information is provided on webpage).

Contents

Figures and Tables

Summary

This report surveys and provides lessons learned on modeling and operations analysis (OA) in Operations Enduring Freedom (OEF) and Iraqi Freedom (OIF). Operations and environmental complexities in Iraq and Afghanistan placed heavy demands on U.S. military commanders, requiring them to make critical, time-sensitive decisions with limited information. Modelers and analysts provided both direct and indirect, or "reach-back," support to commanders in both theaters to help them make well-informed decisions across the full spectrum of counterinsurgency and irregular warfare (COIN and IW) operations. Based on our analysis of both the literature and interviews with commanders and analysts, we identified four general categories that encompassed most decision support: (1) force protection; (2) logistics; (3) campaign assessment; and (4) force structure.

We assess that modelers and analysts were able to successfully inform many force protection and logistics decisions, but they were less effective in supporting campaign assessment and force-structuring decisions (each category is described in greater detail below). Scope, scale, complexity, and the opacity of the operational environment were critical variables: Modeling and analysis were effective at addressing difficult but relatively concrete tactical and operational problems, but less useful in addressing complex strategic problems that required detailed analysis of the operational environment. This finding was perhaps unsurprising given the complexity of the environments and operations in both theaters, but it does help refine understanding of the ways in which modeling, simulation, and analysis might be used most effectively in IW. It also informs decisions on current and future investment in modeling and analytic capabilities. The theory and practical methods for force protection and logistics decision support are generally sound and warrant continued investment, while the theory and methods that support campaign assessment and force employment would benefit from additional investment in theoretical research rather than in applications.

Our research focused on four tasks: (1) identify decisions in campaigns such as OEF and OIF that could benefit from modeling, simulation, and analysis; (2) review and assess the ways in which analysts have attempted to address these decisions; (3) develop insight into the challenges they faced; and (4) find ways in which modeling, simulation, and analysis might be improved. To accomplish these tasks, we conducted a detailed analysis of existing literature on modeling, simulation, and analysis for COIN specifically, and for IW more broadly. We also interviewed 115 commanders and analysts who had experience in Afghanistan, Iraq, or in both theaters.

Lessons Across Four Categories of COIN and IW Decision Support

Four chapters in the main report are devoted to discussing examples (in the form of vignettes in some cases) of how modeling and simulation (M&S) or analysis helped support commanders' decisions. Each chapter describes commanders' insights into decisions within each category, and each contains vignettes describing modeling and analytic support to those decisions.

1. **Force Protection** encompasses all efforts to reduce casualties and damage to friendly forces—including armor improvements, electronic countermeasures, and active efforts to eliminate enemy forces before they can attack. Most commanders interviewed stated that they most needed and most benefited from counter–improvised explosive device (C-IED) decision support in Afghanistan and Iraq. Countering direct and indirect fire mattered, but these were rarely cited as a major concern. Modeling and OA techniques modified from previous work, as well as those developed specifically for C-IED analysis, saved countless lives, preserved millions of dollars in material, and played a significant role in increasing the freedom of movement of U.S. forces in Afghanistan and Iraq.

2. **Logistics** decision support occurred at all levels and ranged from simple tactical calculations to theater-level modeling. Commanders and analysts described a broad range of logistics-related efforts, including those designed to control the movement of supplies, to find efficiencies in aerial transport deployments, and to optimize the location of specialty surgical teams. The tactical, operational, and strategic cases we examined were far more amenable to traditional operational analysis techniques (including modeling) than campaign assessment or force structuring. This stems in great part from the fact that analysts supporting logistics decisions relied primarily on readily available Blue force data and less on complex and often inaccurate and incomplete environmental data such as the number of insurgents or the true character of popular sentiment. The problems that logistics analysts faced were difficult but often straightforward. We identify specific cases in which logistics modeling and analysis saved money and directly reduced threat of injury or death to U.S. military personnel.

3. **Campaign Assessment** is the commander's effort to determine progress against mission objectives in order to optimize planning and resource allocation. Assessments are tools that support a range of decisions, such as how to allocate forces, when to change strategy, and when to request additional support. Yet campaign

assessment offers few opportunities to link analytic effort with operational outcomes. Interviewees reported frustration associated with a poorly defined problem, inadequate data, and a lack of common, validated methods. Commanders had a hard time articulating their requirements, and analysts had a difficult time trying to support decisionmaking. Further, confusion over the definitions and purposes of *analysis* and *assessment*—an issue we address in the report—further undermined analytic support to campaign assessment efforts. Our review of the literature and our interviews did not reveal any clear campaign assessment successes for OEF or OIF.

4. **Force Structuring** decisions encompass the determination of force requirements, force shaping, and force employment. In other words, how many and what kind of troops are needed, and how should they be used? Commanders are responsible for providing policymakers with a clear rationale for their force-structuring requirements, but the analytic community has not yet provided them with a methodology that provides a clear rationale for COIN and IW force-structuring requirements. In general, commanders and analysts have taken one or more of three approaches in an attempt to determine strategic force requirements for COIN: (1) troop or force ration calculations; (2) troop density calculations; and (3) troop-to-task calculations. However, none of these approaches were considered generally sound and effective by policymakers, commanders, or the analytic community. This leaves a critical gap in IW decisionmaking.

Resource Allocation

Where should the Department of Defense (DoD) invest to close the M&S gaps? In this report, we show that modeling, simulation, and analysis have proven clearly useful to support two of the four categories we addressed: force protection and logistics (Chapters Three and Four). We were unable to identify any clear successes from among the various efforts to support campaign assessment and force structuring presented in Chapters Five and Six. Hundreds of military personnel and civilians have worked to develop and improve modeling, simulations, and analytic tools and methods to support decisions across all four categories between 2001 and 2012. It is likely that DoD has invested considerable sums in these developmental efforts over the past 11 years. We are not aware of any cost-benefit analyses conducted to determine which of these specific efforts bore fruit and therefore represented good investments. Based on the information provided to the authors for this report, it appears that DoD would more likely achieve immediate, practical success by investing in force protection and logistics methods and tools rather than in IW campaign assessment and IW force-structuring employment methods and tools.

This apparent difference in opportunity presents a dilemma for those considering future investment: Should DoD weight its investment toward those models, simulations, and analytic tools and methods that have already proven to be useful; should it attempt to address what appears to be a capabilities gap by focusing investment toward these gaps; or should it spread its assets to achieve some kind of parity? We do not weight the value of the various decisions that commanders face in COIN and IW—there is no evidence to show that campaign assessment or force structuring are more or less important than force protection or logistics in any one campaign or across recent COIN and IW cases. Failing to invest in any one category might be equated to a failure to support a selected range of commanders' COIN and IW decisions. However, given that the latter two are least amenable to the kinds of quantitative models and tools typically associated with M&S and operational analysis, we argue that further investment in structured techniques of similar purpose and type—*without a reconsideration of assessment and force-structuring theory*—is putting good money after bad.

We make the following resource allocation recommendations:

- **Leverage existing gap identification to help allocate investment.** DoD should reexamine the gaps in decision support identified in this report in order to better allocate M&S investment.
- **Invest selectively in campaign and strategic assessment and in force structuring.** DoD should continue to diversify investment across all four categories of support covered in this report. However, it should invest more selectively in the development of campaign assessment and force-structuring methods. More is not necessarily better for these two problem sets, at least not until the analytic community resolves the issues with theory, methods, and data described in this report.
- **Invest in efforts to help identify promising methods.** In the near term, the best investment in campaign assessment and force-structuring support is to help the community of experts—military staffs and commanders, operations research systems analysts (ORSAs), social scientists, modelers, intelligence professionals, and general researchers—discriminate between the broad categories of approaches and methods that are likely to provide effective decision support and those that are not. Once these issues have been resolved, investment can be directed to address individual methods and tools with greater confidence.

Each investment in modeling, simulation, and analysis should be predicated on the understanding that the COIN and IW mission and environment places restraints on applicability of many commonly used methods, particularly on those that require large quantities of data.

Additional Findings

In addition to identifying issues associated with supporting different types of decisions, we also identified a range of other findings.

1. **Most decision support derives from simple analyses, not complex modeling.** This is true even while DoD and the supporting community strive to develop models and simulations in support of IW.
2. **Reachback support for COIN and IW is useful, but its role is limited.** Many commanders and analysts praised the role of reachback support for OEF and OIF, but most also noted that this support is bounded by the timeliness and remoteness of operations.
3. **Data quality for many types of data in COIN and IW is generally poor and inconsistent.** In OEF and OIF, data were generally incomplete, inaccurate, and inconsistent. Data-quality issues were sometimes manageable at the tactical level, but rarely at the strategic level.
4. **There is no clear understanding of what is meant by analysis or assessment in COIN and IW.** DoD provides a range of complex, overlapping, and sometimes-contradictory definitions for these terms, and the lack of clear delineation between the two often led to confusion and sometimes to misallocated resources in OEF and OIF.
5. **Some commanders were insufficiently prepared to use analysts or read analyses.** In many cases, OEF and OIF presented commanders with their first introduction to analysts and analyses. Many commanders were not prepared to optimize the use of their trained analysts, and could not read their analyses with a sufficiently informed and critical eye.
6. **Simulation, or wargaming, is useful for decision support in COIN and IW but has limits.** Simulation has helped analysts think through complex problems and helped prepare commanders and staffs for deployment to OIF and OEF. However, the complexity of the IW environment and the lack of good, consistent data preclude the use of simulation as an effective real-world, real-time decision support tool at the operational level (e.g., regiment or brigade) and above.

Acknowledgments

We would like to thank LTC Scott Mann, USA (Ret.) for facilitating our access to special operations commanders. The research team was granted to exceptional access to analysts at the TRADOC Analysis Center thanks to its then-director, Michael Bauman. The same was true for the Army's center for Army Analysis. Its director, Ed Vandiver provided the research team unlimited access to his staff. We thank both Mr. Bauman and Mr. Vandiver for their support, as well as the analysts we interviewed for taking the time to meet with the study team. We also thank all of the commanders, analysts, and others who took time from their schedules to provide us with crucial interviews. Finally, we express our gratitude to the reviewers of this report: RAND colleague John Matsumura and Jon Schroden at the Center for Naval Analyses. Their comments and suggestions greatly improved the quality of the report.

Abbreviations

AAG	Afghan Assessment Group
ABRCS	Agent-Based Rational Choice Stakeholder
AHS	Actionable Hot Spot
ANSF	Afghan National Security Force
AoR	area of responsibility
APL	Applied Physics Laboratory
ASCOPE	Areas, Structures, Capabilities, Organizations, People, Events
AtN	Attack the Network
BCT	Brigade Combat Team
BFT	Blue Force Tracker
CAA	Concepts Analysis Agency
C-IED	counter–improvised explosive devices
CMO	civil-military operations
CNA	Center for Naval Analyses
COA	course of action
COAG	Core Operational Analysis Group
COIN	counterinsurgency
COP	Command Observation Post
DF	direct fire
DIME	Diplomatic, Informational, Military, Economic
DoD	Department of Defense
FOB	forward operating bases
HVI	high-value individual
IBCT	Infantry Brigade Combat Team
IDC	Information Dominance Center
IDF	indirect fire

IED	improvised explosive device
IJC	ISAF Joint Command
IO	information operations
ISAF	International Security Assistance Force
ISR	intelligence, surveillance, and reconnaissance
IW	irregular warfare
IWTWG	Irregular Warfare Tactical Wargame
JCATS	Joint Conflict and Tactical Simulation
JIEDDO	Joint IED Defeat Organization
JNEM	Joint Non-Kinetic Effects Model
JPL	Jet Propulsion Laboratory
KLE	Key Leader Engagement
KTD	key terrain districts
LOO	line of operation
M&S	modeling and simulation
MASHPAT	Marine Assault Support Helicopter Planning Assistance Tool
MISO	military information support operations
MMT	methods, models, and tools
MOE	measure of effectiveness
MOP	measure of performance
MORS	Military Operations Research Society
MSR	main supply route
NATO	North Atlantic Treaty Organization
NDU	National Defense University
OA	operations analysis
OEF	Operation Enduring Freedom
OIF	Operation Iraqi Freedom
OND	Operation New Dawn
OR	operations research
ORSA	Operations Research Systems Analysis
OSD-CAPE	Office of the Secretary of Defense, Cost Assessment and Program Evaluation
PIR	passive-infrared

PMESII-PT	Political, Military, Economic, Social, Information, Infrastructure, Physical environment, Time
PSO	peace support operations
PSOM	Peace Support Operations Model
RC	regional command
RCP	route clearance package
RSTA	reconnaissance, surveillance, and target acquisition
SFA	security force assistance
SIGACT	significant activity
SIP	Strategic Interaction Process
SME	subject-matter expert
SoI/CLC	Sons of Iraq/Concerned Local Citizen
TRAC	TRADOC Analysis Center
TRADOC	Training and Doctrine Command
TTP	tactics, techniques, and procedures
UAS	unmanned aerial systems
UK	United Kingdom
UK MoD	United Kingdom Ministry of Defence
WAD	Warfighting Analysis Division (J-8)

CHAPTER ONE

Introduction

This report provides research findings intended to identify lessons learned from the use of modeling, simulation, and operations analysis (OA) in support of commanders' decisions in Operation Enduring Freedom (OEF) and Operation Iraqi Freedom (OIF).[1] Our findings contribute to answering three questions:[2]

1.   How did military and civilian analysts support decisionmaking in these large-scale counterinsurgency (COIN) campaigns?
2.   How effective were these efforts in improving the quality of commanders' decisions, thereby supporting U.S. strategic goals?
3.   How could modeling, simulations, and OA be improved to better support future COIN and, more broadly, irregular warfare (IW) operations?[3]

Modeling, simulation, and OA as well as systems analysis have routinely been used to support course-of-action selection in conventional warfighting. For centuries, militaries the world over modeled various aspects of siege engineering, and Allied forces applied Lanchester differential equations to support force employment analysis in World War I.[4] Application of operations research (OR) modeling for military application was formalized during World War II within both the United Kingdom (UK)

---

[1]   We also address Operation New Dawn (OND) but focus on OEF and OIF.

[2]   We recognize that the U.S. Air Force and Navy as well as nonservice entities played a significant role in providing modeling, simulation, and analysis support to OEF and OIF. However, we focus specifically on the two military services that were most heavily involved in prosecuting the wars in Afghanistan and Iraq: the U.S. Army and Marine Corps. While we do not directly address intelligence analysis methods or efforts, we recognize that increasing data and staff integration means that the lines between operations analysis and intelligence analysis have blurred. Many of the methods we describe in this report depend on commonly available intelligence information reporting.

[3]   We address the differences between COIN and IW below and explain how our findings on COIN should and should not be extrapolated to IW more broadly.

[4]   A U.S. Army history of OR notes that the use of OR-like methods dates back to approximately 3200 B.C. See Army Logistics University, *Operations Research/Systems Analysis (ORSA) Fundamental Principles, Techniques, and Applications*, ORSA Committee, October 2011, p. 2.

and U.S. militaries. The first computerized simulation of conventional combat was the "Air Defense Simulation," developed by the Army Operations Research Office at Johns Hopkins University in 1948. This was followed by the "Carmonette" series of simulations in 1953, which represented ground combat at the levels of the individual soldier and company.[5] Since then, the number and sophistication of these simulations has grown hand-in-hand with the rapid expansion of computing power.[6]

Similarly, the use of standardized and ad hoc OA tools and techniques has grown, and the use of OA is now common at staffs down to the brigade (Army) or regimental (Marine) level. Commanders faced with the complexities of Afghanistan and Iraq operations have become increasingly dependent on analytic support to help them make sound decisions that will enhance their operations or, at the very least, will not undermine their long-term objectives. Between 2001 and 2013 it became increasingly common to find operations research systems analysts (ORSAs) at multiple staff levels as the size and complexity of U.S. and coalition forces increased and then rapidly decreased. Similarly, reachback teams of modelers and ORSAs—often one and the same—grew or were created to provide both long-distance support to commanders and also to provide staffs with a pool of capable analysts.

Because the COIN campaigns in both Afghanistan and Iraq were developed mid-stride, after U.S. forces had trained and deployed and as the insurgencies emerged, the modeling, simulation, and OA efforts for COIN were often experimental and ad hoc. Many analysts compare the development and application of their analysis in OEF and OIF during the 2000s to building an airplane in midflight.[7] While some theories, methods, and associated technologies advanced in leaps and bounds, others lagged or received insufficient investment or attention. And because COIN doctrine was also being built in midflight—the U.S. manual for COIN was not published until 2006— the analysts trying to adapt or develop new methods did so with only a limited understanding of what an optimized COIN campaign should look like. We explore these issues and many others, provide a range of vignettes to exemplify both successes and challenges, and offer recommendations for improvement. It is first necessary to establish terms.

[5]   Carrie McLeroy, "History of Military Gaming," *Soldiers Magazine*, August 27, 2008.

[6]   Charles R. Shrader, *History of Operations Research in the United States Army*, Vol. 1: *1942–62*, Washington, D.C.: Office of the Deputy Under Secretary of the Army for Operations Research, U.S. Army, Center for Military History publication 70-102-1, August 11, 2006. This history of operations analysis in the U.S. Army tracks many of these developments from World War II through 1973. The entire history of modeling, simulation, and analysis in support of warfighting is expansive and beyond the scope of this report. The References section offers a number of excellent resources that describe various parts of this history.

[7]   This observation is based partly on interviews, partly from the literature, but primarily from extended off-the-record conversations between more than 100 ORSAs and members of the RAND research team between 2004 and 2013.

Clarifying Terminology

OA is generally understood to be the application of common theories, tools, and techniques for the purposes of improving or optimizing an organization's capabilities and performance.[8] In the Department of Defense (DoD), trained and educated ORSAs generally conduct OA, although non-ORSA analysts use some OA techniques.[9] Modeling and simulation (M&S) tend to be spoken of as a single entity—or at least equivalent entities. This leads us to definitions we will use for the purposes of this report. Chapter Two provides examples of other versions of these definitions to demonstrate the wide range of interpretations that affect the application of modeling, simulation, and analysis.

The three processes are closely intertwined: One cannot develop a model without conducting analysis, even if analysis in this sense is no more than critical thinking, and one cannot develop or run a simulation without building at least a rudimentary conceptual model.

Use of the Term *Model* in This Report

The official DoD definition of a model is "a physical, mathematical, or otherwise logical representation of a system, entity, phenomenon, or process."[10] We view units conducting IW operations to constitute a system. The time components account for the evolving nature of IW operations and the space component accounts for the venue. However, the real value a model offers is insight. Analysts and operators report that they gained considerable understanding of their IW operations simply through the process of developing the model.

As a representation of a system, a model focuses on the component parts of the system and how they relate to each other. For example, a computer model of conventional combat between two adversaries might consist of several modules, such as ground operations, air defense, air-to-air and air-to-ground operations, command and control, and Intelligence, Surveillance and Reconnaissance (ISR). These modules do not operate independently over time so the relationship among the modules is part of the model as well.

---

[8] This definition of OA is drawn from multiple sources, discussions with ORSAs, and peer review comments.

[9] The difference between training and education for operations analysis is a subtle but important distinction. Many ORSAs receive formal education at the master's level to qualify as military operations researchers. There are also a number of short training courses in specific techniques offered to ORSAs and others. Most ORSAs have some mix of formal education and specific training.

[10] DoD, *DoD Modeling and Simulation (M&S) Glossary*, DOD 5000.59-M, Under Secretary of Defense for Acquisition Technology (USD[AT&L]), January 1998. This glossary has been updated: DoD Directive: *DoD Modeling and Simulation (M&S) Management*, Number 5000.59, USD(AT&L), August 8, 2007. The updated version references the older document for the definition of "model."

Usually, the exact relationships among the variables in complex models such as those describing IW operations are not known. What is known, in many cases, are the influences the variables have on each other. For example, a sensor detection of an enemy convoy increases the command's situational awareness. Because the exact relationship is not known (how many "units" of situational awareness derive from a single "unit" of sensor detection), we generally deal in likelihoods or probabilities. A second influence (and one that is potentially more relevant to IW and COIN operations) is that of human behavior on operations. In this realm, there is knowledge of influence direction but often little or no knowledge of precise mathematical relationships and physics formulas that can be drawn upon.

All simulations employ models but not all modeling and use of models involves simulation as it is commonly understood. Simulation is just one analytic technique that involves the use of models. When the relationship among the variables are known, a simple closed form solution is available. For example, Newton's second law of motion is a model of the relationship between the mass of an object, its acceleration, and the force exerted on it. The relation is described by the equation $F=Ma$, where F is force exerted on the object, M is its mass, and a is its acceleration.[11] This is referred to as an *analytic solution* model; i.e., the relationships among the model variables are known exactly and therefore produce an exact solution analytically. This is a case where causality is proscribed. But these rarely exist in military operations, except for routine logistics and manpower models. What we experience instead are correlations; for example, support for the coalition decreases as civilian casualties increase. However, other factors may be in play and therefore we cannot state that civilian casualties *cause* decreases in support. The inability to identify cause and effect, or even a compelling argument for meaningful correlation in many cases, undermines the utility of modeling for the more complex key decisions: campaign assessment and force structuring.

Use of the Term *Simulation* in This Report

The official DoD definition of a simulation is "a method for implementing a model over time."[12] In effect, simulation is the *imitation* of the operation of a real-world process or *system* over time. A useful definition we have adopted for this report is: *Simulation is the manipulation of a model (usually using a computer) in such a way that it operates on time or space to gain some understanding of how the modeled system operates.*[13]

The importance of simulation is that it provides understanding of how the real-world system actually works, as represented by the model. This is critical when simulating COIN or IW operations. Gaining such an understanding is central to the use-

---

[11] Benjamin Crowell, *Light and Matter*, Creative Commons Attributions, 2011, Section 4.3.

[12] DoD, 1998; DoD Directive, 2007.

[13] For a fuller discussion of simulation see Averill M. Law and W. David Kelton, *Simulation Modeling and Analysis*, 1st ed., McGraw-Hill Higher Education, 1982, Chapter 1.

fulness of models and their use in simulation to inform critical decisions. However, the simulation process is only as good as the model created—and here is the principal difficulty with M&S in complex operations such as COIN or IW.

In environments like Afghanistan or Iraq, the logical relationships among the various components are not always known. For example, if we are to create a COIN model using the system-of-systems theory depicted in joint doctrine, the components might consist of the insurgents, the host-nation forces, the allied forces, and the population.[14] This last component might be further divided by demographics, ethnicity, tribal affiliations, and perhaps religion. Clearly, all of these components are related in some way, but how? In many cases, it is also not possible to segregate components into neat subsystems: If a government official is also an insurgent, that individual exists simultaneously within two subsystems. A simulation of a model created without a reasonable understanding of these relationships, or of the complexities inherent in individuals' and groups' identities and roles, may lead to erroneous conclusions.

As in all applications, effective M&S support for critical decisions in COIN and IW depends upon how well the model represents the environment being simulated. Creating a model of a complex environment such as Afghanistan or Iraq is problematic because of the extreme uncertainties in our understanding of the interactions among the various system components. Using subject-matter experts (SMEs) to take on the role of the components in a human-in-the-loop simulation—as with peace support operations models (PSOMs)—appears to be a popular approach.

Defining and Differentiating Between "Analysis" and "Assessment"

In order to understand how modelers and analysts supported commanders in Afghanistan and Iraq, it is necessary to briefly examine the differences and overlap between the terms *analysis* and *assessment*. This is particularly relevant because we identify one of the four primary fields of support as campaign *assessment*. DoD does not clearly define OA, but does define OR, which it conflates with OA. OR is "the analytical study of military problems undertaken to provide responsible commanders and staff agencies with a scientific basis for decision on action to improve military operations." Sometimes ORSAs and others refer to their work as assessment, but there are at least semantic differences between analysis and assessment. DoD undertakes a wide array of assessments across all of its component services, agencies, and functions. It defines the term *assessment* as:[15]

---

[14] Nearly all capstone U.S. joint doctrinal publications include a presentation and description of the System-of-Systems Analysis modeling theory. For a discussion of this approach, see Ben Connable, *Embracing the Fog of War: Assessment and Metrics in Counterinsurgency*, Santa Monica, Calif.: RAND Corporation, MG-1086-DOD, 2012, pp. 286–287.

[15] U.S. Joint Chiefs of Staff, *Department of Defense Dictionary of Military and Associated Terms*, Joint Publication 1-02, Washington, D.C., as amended through May 15, 2011, p. 21.

1. A continuous process that measures the overall effectiveness of employing joint force capabilities during military operations.
2. Determination of the progress toward accomplishing a task, creating a condition, or achieving an objective.
3. Analysis of the security, effectiveness, and potential of an existing or planned intelligence activity.
4. Judgment of the motives, qualifications, and characteristics of present or prospective employees or "agents."

DoD definitions of analysis and assessment overlap, are not sufficiently descriptive, and are somewhat confusing. For example, the definition of assessment describes four separate efforts that are not clearly linked: (1) a *part* of campaign assessment (daily information monitoring); (2) the longer-term *process* of campaign assessment that *includes* monitoring; (3) an examination of intelligence analytic methods; and (4) intelligence agent or source screening. The last two are distinct from the first two and also clearly distinct from each other. To further confuse matters, by these definitions, assessment is often an overarching process that might or might not include analysis, and analysis might in some cases constitute an assessment. We describe how this ambiguity affected analytic support in OEF and OIF, but it is possible to place these two efforts—analysis and assessment—in broad bins.

Based on the existing literature, interviews for this report, and ongoing research of this issue by two of the lead authors, it is possible to draw a clearer distinction between analysis and assessment. Perhaps the best way is by viewing analysis as a focused process that uses an identified method, and assessment as a decision support function that might *or might not* include one or more analytic processes and methods. For example, a military commander in Afghanistan can write a campaign assessment based only on his personal observations and understanding of the campaign, or he can incorporate a time-series analysis of violent incidents over time, which requires the focused application of a method. In another example, a staff officer can present an analysis of convoy movement as a stand-alone assessment of transportation efficiency—in this case, the analysis serves as an assessment. This distinction and these examples still leave room for what may be unavoidable ambiguity.

The various types of analysis used in OEF and OIF are somewhat easier to define and separate. The U.S. Army Logistics University describes the analytic process for ORSAs. This ten-part process is depicted in Figure 1.1. It shows a logical, cyclic, scientific progression that in theory would be well suited to provide effective decision support in almost any situation. Interviews for this report and also existing interviews with deployed analysts revealed a fairly consistent effort by most ORSAs to follow this logical, scientific model—or at least the parts of it that were relevant to each real-world

Figure 1.1
ORSA Scientific Approach to Problem-Solving

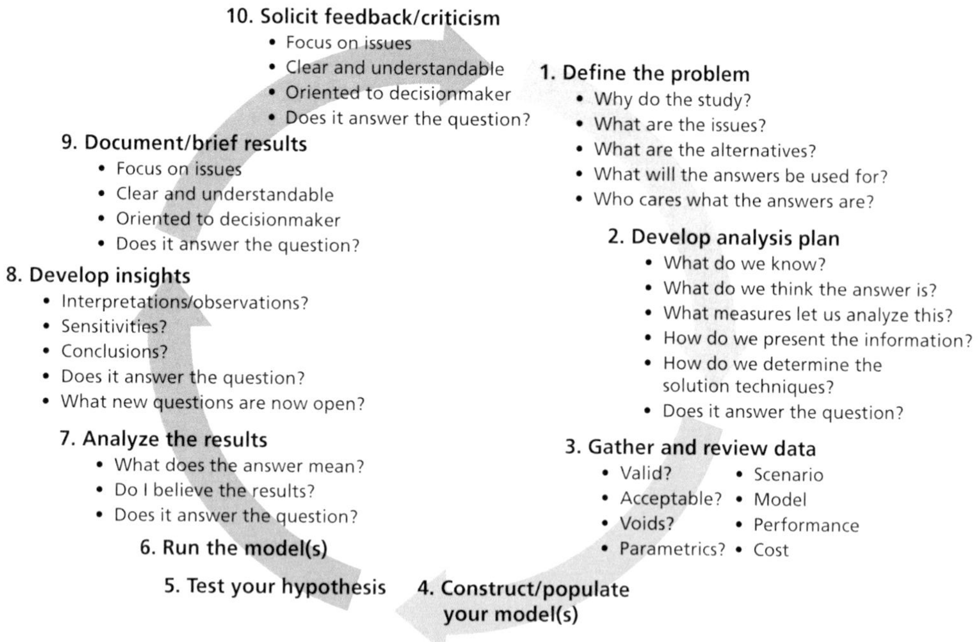

10. Solicit feedback/criticism
- Focus on issues
- Clear and understandable
- Oriented to decisionmaker
- Does it answer the question?

9. Document/brief results
- Focus on issues
- Clear and understandable
- Oriented to decisionmaker
- Does it answer the question?

8. Develop insights
- Interpretations/observations?
- Sensitivities?
- Conclusions?
- Does it answer the question?
- What new questions are now open?

7. Analyze the results
- What does the answer mean?
- Do I believe the results?
- Does it answer the question?

6. Run the model(s)

5. Test your hypothesis

1. Define the problem
- Why do the study?
- What are the issues?
- What are the alternatives?
- What will the answers be used for?
- Who cares what the answers are?

2. Develop analysis plan
- What do we know?
- What do we think the answer is?
- What measures let us analyze this?
- How do we present the information?
- How do we determine the solution techniques?
- Does it answer the question?

3. Gather and review data
- Valid?   • Scenario
- Acceptable? • Model
- Voids?   • Performance
- Parametrics? • Cost

4. Construct/populate your model(s)

SOURCE: The Army Logistics University, 2011, p. 5.
RAND RR382-1.1

problem—to support commanders in OEF and OIF.[16] However, these interviews also revealed the ways in which commanders' preferences, combat timelines, resource limitations, the insurgents, and the environment all affected their ability to execute this cycle as intended. It would be useful to consider this cycle as a basis for comparison to the vignettes in the literature review and interviews. We address this cycle again in Chapter Seven.

We refer to analysis here as analytic decision support conducted primarily outside of the intelligence process—this report does not address processes or lessons learned for the intelligence process, except incidentally through individual vignettes. In many instances in both OEF and OIF, OA informs intelligence analysis, and intelligence reporting and analysis also inform further OA. This is especially true in force protection operations. Chapter Three describes a rather large body of analytic support to counter–improvised explosive device (C-IED) operations that focused almost entirely on supporting intelligence analysis. In some cases, roles are mixed. For example, in late 2009 and early 2010, the International Security Assistance Force (ISAF) Joint Command (IJC) staff made the determination to place its campaign assessment staff,

---

[16] For example, not every problem required the development of a model.

consisting of ORSAs and infantry officers, within the intelligence Information Dominance Center. In Iraq and Afghanistan, intelligence analysts at the brigade level and above often use OR-derived charts to support their analyses. In other cases, OR supported intelligence analysis more directly. For example, in 2003 a Concepts Analysis Agency (CAA) ORSA officer deployed in support of the Combined Joint Task Force-7 staff, during which he helped develop the intelligence section's intelligence preparation of the battlefield process, building templates to combine multiple data streams to support target selection.[17] Chapter Two presents some general distinctions between operations and intelligence analyses.[18]

Differences Between Campaign Assessment and Intelligence Analysis

RAND proposed differences between campaign assessment and intelligence analysis in a 2012 report on assessment in COIN as depicted in Table 1.1.

Systems Analysis

DoD does not define the term *systems analysis*, which is also part of the ORSA process. For DoD, systems analysis is most commonly understood to be a process of understanding institutional or service-level program issues, such as budgeting. While some

Table 1.1
Differences Between Campaign Assessment and Intelligence Analysis

Characteristic	Campaign Assessment	Intelligence Analysis
Primary purpose	Assess progress against operational and strategic objectives	Explain behavior and events and predict future behavior and events
Process	Describes and explains progress, recommends shifts in resources, strategy, informs operations	Describes and predicts behavior and actions in the environment, informs courses of action for operations and policy
Method	Any relevant and useful method	All-source analysis using structured, doctrinal methods within prescribed intelligence oversight limits
Sources	Any availiable sources, including friendly operations reports and completed intelligence reports	Limited to examination of enemy, foreign civilian, and environmental information
Creators	Representatives of all military staff sections and military commanders	Trained intelligence analysts
Time frame	Show progress over long periods	Timely, degrades in value over time
Classification	Can be mostly or entirely unclassified	Almost always classified or restricted

SOURCE: Connable, 2012, p. 3.

---

[17] CAA, *Analytic Support to Combat Operations in Iraq (2002–2011)*, Deployed Analyst History Report, Vol. 1, March 2012, p. 29.

[18] For a differentiation between intelligence analysis and operational analysis, see Connable, 2012, pp. 2–4.

parts of force structuring for OEF and OIF might be construed as systems analysis, the work done to understand force-structuring issues was primarily OA. Throughout this report we focus primarily on OA and the work of ORSAs in both deployed and reachback roles. For this report, the term *analysis* generally refers to the ways in which ORSAs used standard and ad hoc tools to provide decision support to commanders. Both OA (or OR) and systems analysis have multiple, conflicting, and contested definitions. For purposes of simplicity, we group the use of operations and systems analysis together under the term *analysis*.[19]

Why Are COIN and IW So Difficult to Analyze and Assess?

Just as there are multiple and often conflicting definitions for OA, OR, and systems analysis, the definitions of COIN and IW are also problematic. As of mid-2013, the simplest DoD definition of COIN is "comprehensive civilian and military efforts taken to defeat an insurgency and to address any core grievances," while the simplest definition of IW is "a violent struggle among state and non-state actors for legitimacy and influence over the relevant population(s)."[20] Using these definitions, IW describes a very broad palette of civil disorder conditions, including civil war, rebellion, insurgency, separatism, and others; each of these definitions is also contested and all overlap somewhat. COIN, on the other hand, is the purposeful effort to defeat an insurgency, which is one of the possible groups fighting in an irregular war. COIN, then, is an action to address one specific form of IW. Therefore, while lessons from COIN may not be universally applicable to all IW situations, the lessons from those efforts in Afghanistan and Iraq are *relevant* to IW more broadly.[21]

The definition of IW also proposes that "irregular warfare favors indirect and asymmetric approaches, though it may employ the full range of military and other capacities, in order to erode an adversary's power, influence, and will."[22] Knowing that irregular—or, for our purposes, insurgent—forces operate asymmetrically and focus their efforts against nonmaterial critical capabilities such as power, influence, and will is important to understanding the challenges posed to M&S in this environment.

For example, during WWII, Army operations analysts were able to clearly and closely track the tonnage of bombs dropped on enemy infrastructure in Germany, determine the effectiveness of attacks, and recommend adjustments to tonnage, flight

---

[19] U.S. Joint Chiefs of Staff, 2011, DoD, 1998, and DoD Directive, 2007, do not contain a definition of systems analysis.

[20] U.S. Joint Chiefs of Staff, 2011, pp. 64 and 106.

[21] For example, see DoD, *Sustaining U.S. Global Leadership: Priorities for 21st Century Defense*, January 2012.

[22] U.S. Joint Chiefs of Staff, 2011, p. 166.

performance, and even pilot training.[23] While no problem in warfare is straightforward, problems like battle damage assessment in a conventional fight present challenges that nearly any analyst and commander can easily grasp. Analysts now have common and well-tested sets of M&S tools to help them understand complex but digestible problems such as conventional targeting performance. By comparison, consider a deployed analyst attempting to understand why a group of insurgents—a group he cannot see, count, or even distinguish from the population—is motivated to fight against the Iraqi or Afghan government. Few, if any, of the tools or analytic concepts that proved so useful in ferreting out more effective bombing methods can help the IW analyst understand the mind of the insurgent.

Previous RAND research revealed many of the challenges inherent in the IW environment. Because operations like COIN are conducted in the midst of large, socio-culturally complex populations, and because the enemy is difficult to identify and understand, the environmental feedback (or data) necessary for sound analysis is often inaccurate and incomplete.[24] The task of understanding the dynamics of a large, often heterogeneous population wracked by violence and displacement is the most difficult part of modeling and analyzing COIN and IW. A common complaint among analysts was the lack of accurate and complete data, even for some of the relatively straightforward OA tasks like convoy optimization. One could argue that while all warfare is complex and chaotic, the degrees of complexity and chaos of COIN and IW make the task of analysis in these operations more difficult than in conventional combat.

The anecdotal evidence presented in this report reveals analysts' daily struggles to understand the complex COIN and IW environments. Each vignette describes efforts to frame useful questions for commanders, apply existing tools to uncommon problems, and find new analytic tools and methods that might better match the kinds of challenges presented by insurgency and the complex and often chaotic environments of Iraq, Afghanistan, and the outer Philippine islands.

Research Objectives and Limits

The purpose of this research was threefold:

1. Identify decisions at all levels of IW that could benefit from more extensive and rigorous modeling, simulation, and analysis.
2. Identify ways in which analysts have attempted to address these decisions and describe a selection of the models and tools they employed.

---

[23] See Shrader, 2006, pp. 30–31.

[24] See Connable, 2012, Chapter 1.

3. Provide insight into the challenges they faced, and suggest ways in which the application of modeling, simulation, and analysis might be improved for current and future operations.

Focus on Operations for the Army and Marine Corps

This report focuses specifically on military ground commanders' decisions and on the support provided to those decisions. Because it is focused on U.S. Army—and to a lesser extent, Marine Corps—experiences, it does not address all of the modeling, simulation, and analytic efforts undertaken on behalf of other services or other elements of the joint community. We note that the U.S. Air Force, the U.S. Navy, and a range of special operations forces have benefited from modeling, simulation, and analysis, and that these efforts have in turn supported decisions at all levels of effort in both theaters. And while this report does address some aspects of intelligence analysis, it does so only peripherally; an examination of modeling, simulation, and analysis within the Intelligence Community would demand a separate research effort. We define some of the distinctions between OA and intelligence analysis herein.

Methodology

We addressed the research objectives along three complementary lines of effort. First, we conducted a literature review that included an examination of existing lessons learned reports on M&S for IW. This effort helped us to identify existing research, locate potential interviewees, and help bound the final report. We expanded this literature review to a broader review of existing approaches and methods, and provide a limited analysis of recorded efforts. Second, we conducted semi-structured interviews with 115 subjects, 82 of whom were ORSAs or other analysts who had either deployed in support of OIF and OEF or provided reachback support to OIF, OEF, or both. The analysts who had deployed generally had been attached as individual augments to a brigade, division, or theater staff, and several analysts interviewed had served at more than one level of command and/or in both OIF and OEF. Of the 115 subjects interviewed, 33 were commanders from both conventional and special operations forces who had commanded at tactical, operational, and strategic-theater levels of authority. Table 1.2 shows both the branch and echelon of operation for each of the commanders interviewed. These interviews were the primary source for our findings, and they provided exemplary vignettes. Third, we conducted a review of a nonrandom representative sample of existing models that had been used in support of modern IW campaigns. Taken together, these three approaches were sufficient to support robust findings and some lessons learned recommendations.

Table 1.2
Branch and Echelon of Commanders Interviewed for This Report

Branch	Echelon						
	Company	Battalion	Brigade	Regional Command	Theater	Global	Grand Total
Aviation		1					1
Civil Affairs			1				1
Engineering		1					1
Infantry	1	6	2		3		13
Intelligence					2	1	3
Logistics		1		1			2
National Guard	1						1
SEALs	1				2	1	4
Special Forces				4	3		7
Grand Total	3	9	3	5	10	2	33

About This Report

This report is divided into eight chapters and one appendix. Chapter Two presents the literature review, which both samples existing literature and provides limited analysis. Findings in Chapter Two are derived solely from the literature review and are intended to reveal existing evidence, ongoing debates, and expert positions on the issues relevant to our analysis. Chapters Three through Six address four categories of key IW decisions: force protection, logistics, campaign assessment, and force structuring. Chapters Four, Five and Six are structured similarly, providing an overview of a specific set of decisions, a description of how commanders interviewed tended to view the associated IW problems, and a series of vignettes drawn from the interviews and documents. We depart from this format somewhat in Chapter Three dealing with force protection. However, the key elements—key decisions, the commanders' view of the problem, and examples of analytic and/or M&S support to key decisions—are addressed. Each of these four chapters is of differing length due to the varying quality and availability of information drawn from the interviews. Chapter Seven presents our overall findings and lessons learned recommendations. The appendix provides a review of selected models and simulations in use in the community. An extensive bibliography identifies sources that informed this report as well as previous RAND reports on modeling, simulation, and analysis for IW.

CHAPTER TWO

Decision Issues and Analysis in COIN and IW Literature

This chapter provides analysis of selected literature on the use of modeling, simulation, and analysis in COIN and IW. The Bibliography contains the full scope of references identified for this report. Observations and findings in this chapter are derived primarily from the literature review; except where explicitly noted, they do not reflect our overarching research and interviews. These observations and findings are intended to set a baseline for the interview portion of this report. We selected literature from existing RAND reports and databases, from a search of professional journals through subscription services, from recommendations provided by experts we interviewed, and from reports provided to us by other military commanders and both military and civilian analysts. This is a review of selected literature and therefore many publications, unofficial reports, and official reports are not included. Because our report is unclassified, it necessarily cannot include classified titles. However, we did seek to obtain unclassified versions of relevant classified reports or briefings whenever possible.

We begin by describing some analysis literature—the Training and Doctrine Command (TRADOC) Analysis Center (TRAC) Irregular Warfare Working Group Report and a CAA report on deployed analytic support to OIF—that specifically addresses the research objectives of this report. The following section explores commanders' decisions in COIN as they appear in the sampled literature. Next, we examine the ways in which modeling, simulation, and analysis are described in the sampled literature, and analyze the ways in which decisions have been addressed by various approaches and methods. This section also addresses observations made by operations researchers at Military Operations Research Society (MORS) conferences on analytic support to COIN and IW. The conclusion summarizes decisions and lessons identified in the literature.

Two Highly Relevant Publications

Two publications, the TRAC "Irregular Warfare Methods, Models, and Analysis Working Group Final Report" briefing and the CAA *Analytic Support to Combat Operations in Iraq (2002–2011)* report, provide detailed insight into commanders' key decisions and also into the approaches, methods, and tools used to support those decisions.[1] We briefly summarize each of these documents, which collectively represent nearly 350 pages of empirically derived gap analysis and rich anecdotal evidence. CAA also plans to publish a deployed analyst report to address its support to OEF.

TRAC Irregular Warfare Working Group Report

The TRAC briefing presents the findings from a working group the center sponsored on methods, modeling, and analysis in support of IW. This working group had three purposes: (1) determine the analytic community's ability to support IW decisions; (2) identify gaps in DoD IW analysis; and (3) recommend solutions for mitigating these gaps over time. The working group included representation from across the military OA community and focused on OA rather than intelligence analysis. Its basis of analysis is centered on U.S. Army warfighting activities, but is also intended to support U.S. Marine Corps IW analysis. The working group applied several empirical methods, each of which is described in the report.[2] The report's key findings are:

1. There are 56 analytic capabilities that might be or have been used to support IW decisions, and there are 35 major gaps across these capabilities in DoD.
2. Of these 35 gaps, 34 are caused or compounded by a lack of credible data.
3. Of these 35 gaps, 20 require what the report describes as "soft science" [social science] expertise to address.
4. Of the 35 gaps, 18 were deemed to be "high-risk" gaps for commanders, and of these 18, 14 fell within the soft-science category.
5. The 20 gaps that require soft-science expertise are gaps in human behavior data.

Figure 2.1 depicts a risk chart taken from the report. It compares the likelihood that a gap in DoD analytic capability will affect an IW decision (along the x-axis, from unlikely to frequent) along with the damage that the lack of analytic capability might cause to a commander's ability to make sound decisions. Gaps with the highest risk are colored in red and occur on the upper right portion of the table. The top nine gaps, identified as representing "extremely high risk," are listed individually below the chart.

The TRAC report goes on to list each of these analytic gaps in detail, linking each to its risk at each level of war (tactical, operational, and strategic), the estimated range of cost to address the gap (e.g., high, low), and what types of support would be needed to address the gap. It specifically identifies the inability of DoD analysis to help commanders understand the population and how to interact with the population in a way that will further campaign objectives. It identified a potentially "catastrophic"

---

[1] Larry Larimer, John Checco, and Jay Persons, "Irregular Warfare Methods, Models, and Analysis Working Group: Final Report," Scripted Brief, TRADOC Analysis Center, TRAC-F-TR-08-035, August 18, 2008; CAA, 2012.

[2] Larimer, Checco, and Persons, 2008, pp. 22–24.

Figure 2.1
Risk Analysis of Decision Analysis Capability Areas

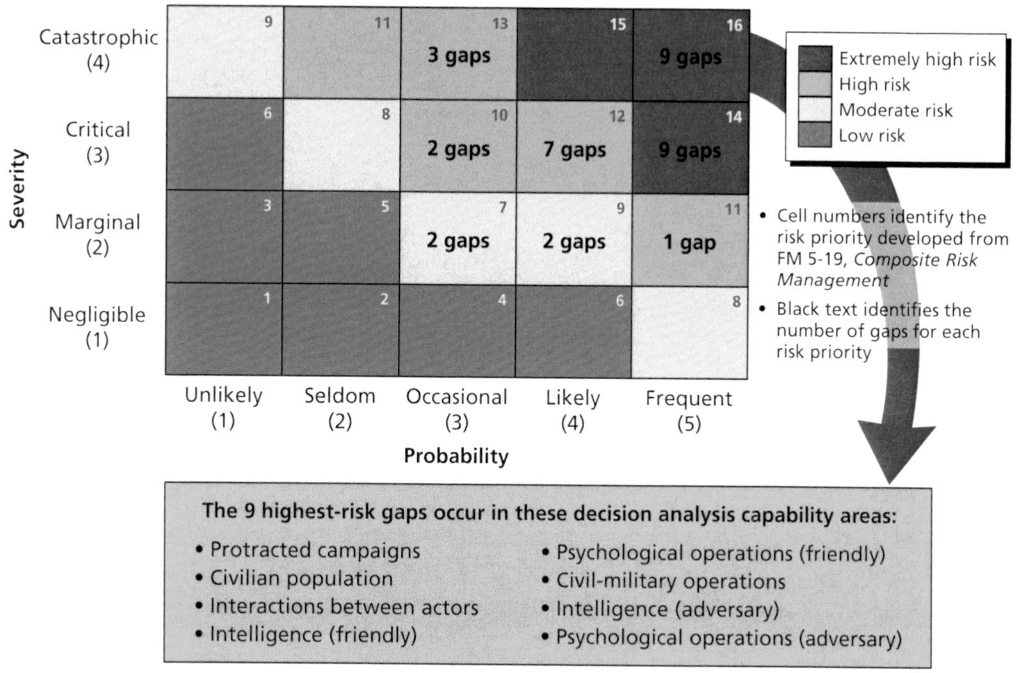

SOURCE: Larimer, Checco, and Persons, 2008, p. 27. The risks were drawn from U.S. Army Field
Manual 5-19, *Composite Risk Management.*
RAND *RR382-2.1*

gap in analytic capability to help commanders understand protracted campaigns and
their "effects over time."[3] The working group also identified more than 100 commanders' IW decisions under 14 separate decision issue categories. These were informed by
both literature studies and interviews with senior leaders, including the Commanding
General of TRADOC, Marine Corps general officers, a Department of State representative, as well as three brigade-level leaders.[4]

Because the TRAC working group was primarily staffed by ORSAs, and focused
on OA issues and methods, it approached the issue of decisions and gaps as primarily issues of data and quantitative method. There appears to be a general underlying
assumption in the working group report that the methods designed to address decisions, and gaps in decision support, would be drawn from the OA toolkit or would be
developed in a way that could incorporate and divine meaning from large quantities
of coded, quantitative data. Therefore, this report reflects a selected set of potential

_____

[3] Larimer, Checco, and Persons, 2008, p. 29.

[4] Larimer, Checco, and Persons, 2008, pp. 43–55.

issues and methods rather than a comprehensive approach. For example, while it does identify the need for "soft science" support, it does not clearly address nonquantitative issues or methods that might be used to address these soft-science issues. It does, however, clearly describe some of the most important issues in COIN and IW analysis, including those pertaining to data and the challenges that the COIN and IW environment presents to operations analysts:[5]

> Data from current operational environments, such as Iraq and Afghanistan, may provide some important data, but much of the data will be limited in its usefulness. In the context of governance and society, the operational environment in Iraq and Afghanistan are different from each other, and will most likely be different from any future operations the U.S. military will be involved in. Even within Iraq the environment is different from one location to the next and differs over time. Effective tactics and techniques in one city may not be effective in another or even in that same city a few months later. For long term analysts, DOD needs to collect behavioral data about potential threat countries from country/regional and social science experts.

The TRAC working group report was released in 2008 and is therefore somewhat dated. However, our literature review and interviews showed that most of the working group's findings remained relevant in 2013.

CAA Deployed Analyst History for OIF

The purpose of the CAA report is to "capture the experience of analysts deployed from [CAA] to Iraq" for OIF and OND.[6] It contains more than 200 pages of anecdotal recounting of individual CAA analysts' deployments derived from internal CAA interviews, examples of the types of decisions addressed at various levels of command, and examples of the analytic methods applied to those decisions. This is a report about the ORSA experience in Iraq, so it is necessarily focused on the application of operations research to support key IW decisions. CAA deployed more than 40 analysts to Iraq between 2003 and 2011, primarily supporting commanders from the brigade to the theater level.[7] The CAA report outlines an impressive array of decisions, analytic approaches, and methods, some of which are described in this chapter and Chapters Three through Six. Analysts supported decisions across almost the entire spectrum of those IW decisions identified in the TRAC IW working group report; the examples are far too numerous to list here. This section presents a selected sample of insights, decisions, approaches, and methods.

---

[5]   Larimer, Checco, and Persons, 2008, p. 34.

[6]   CAA, 2012, p. 2.

[7]   CAA, 2012, pp. 21–22. The first analysts deployed in 2002 to Kuwait in support of preinvasion planning.

Major Stephanie Tutton deployed in support of the theater command for OIF in 2004. She echoed the sentiment of other deployed analysts when she noted that an ORSA deployment was an "ongoing educational process." The purpose, relevance, and effectiveness of an ORSA were dependent on a range of variables, including the competence of the individual analyst and the predilections of the commander. The CAA report describes Tutton's view of the ORSA's role in IW decision support:[8]

> Decision-makers continually repeated a cycle of gathering data, assimilating information and intelligence, conducting operational assessments, making a decision and providing guidance to subordinates and staff. The ability of an ORSA analyst to assist in the warfighter's decision-making process during the rapid pace of combat operations, with a series of decision support tools, in the least obtrusive manner, was an art form in and of itself.

This section goes on to describe the role of the ORSA more specifically:[9]

> ORSA analysts improve information and data quality through collection methods, data mining, and understanding of correlation versus causation and, most importantly, the ability to interpret data and understand its limitations. These capabilities, coupled with the ability to relate operational objectives to desired effects and goals, and to develop consistent and relevant metrics with which to conduct assessments, greatly enhance the decision-making process.

Many of the analysts interviewed noted that deployed analysts must seek out opportunities and self-initiate projects whenever possible.[10] And most of them recognized a need for a common set of tools and capabilities. Competence with the Microsoft Excel and Access tools appears to be a baseline requirement for successful decision support analysis, an observation that coincides with our interview findings on the OR approach to analysis. The CAA interviews also showed an expected bias toward quantitative approaches, methods, and data. For example, one analyst noted that campaign measures of effectiveness (MOEs) "should be quantifiable when feasible."[11] The analysts' stress on quantitative approaches to campaign assessment falls in line with contemporaneous doctrine and other DoD literature on assessment.[12] Analysts' approaches were guided as much by commanders' preference as by their training, education, and personal predilections: Many analysts noted that commanders and staffs *expected* them

---

[8] CAA, 2012, p. 38. These appear to be paraphrases rather than direct quotes.

[9] CAA, 2012, p. 38.

[10] For example, see CAA, 2012, p. 28.

[11] CAA, 2012, p. 28.

[12] For example, see Joint Warfighting Center, U.S. Joint Forces Command, J9, Standing Joint Force Headquarters, *Commander's Handbook for an Effects-Based Approach to Joint Operations*, Suffolk, Va., February 24, 2006a.

to produce quantitative analyses. Two ORSAs who served in Iraq supported key decisions including:

1. When and where can the coalition withdraw forces?
2. How should the coalition best distribute scarce electronic warfare equipment in the C-IED effort?
3. Where should we distribute limited reward/reconstruction funds to improve relations with the populace?

In order to support these decisions, "Commanders and their staffs constantly sought quantifiable, or at least readily comparable, information on which to base these decisions."[13] It is not clear whether the requirement for quantitative analysis derived organically from commanders' needs, or if the commanders became accustomed to receiving quantitative analysis, liked it, and therefore sought more of the same over time.

Analysts supported campaign assessment but were also able to focus on projects that delivered clearer and more readily obtainable results. Most of these fell within the realm of either force protection or logistics support. For example, Major Andrew Farnsler deployed to Iraq in 2006 in support of Multi-National Corps—Iraq, where he performed a broad range of analyses from campaign assessment to logistics optimization and force protection. In one case he was able to help the command improve the performance of its ground-based ISR systems:[14]

> This study determined and measured the success of ISR systems and was the first step to improving their operational and tactical employment. . . . The command defined success as system acquisition resulting in enemy BDA [battle damage assessment] . . . Major Farnsler went to Balad [base] and viewed ISR operations from sensors to integration. He had the opportunity to talk with [ISR] operators, maintenance personnel, and a battalion operations officer, all of whom provided great insight into successful [ISR] use. The Command implemented many of [Farnsler's] recommendations.

CAA analysts also trained nonanalysts in the use of basic analytic tools, executed large database management and data cleaning projects, developed and deployed data visualization tools, and contributed to a range of working groups and staff section efforts as critical thinkers. CAA wrote and disseminated handbooks for deploying ORSAs, a handbook to help commanders employ their ORSAs effectively, and implemented a two-week predeployment analytic training program for its analysts.

---

[13] CAA, 2012, p. 63.

[14] CAA, 2012, p. 69.

Decision Issues Identified in Literature

This section describes many of the commanders' decisions that we identified in the literature, while Chapters Three through Six describe these decisions in more detail by analytic category. The literature review revealed few comprehensive or even specifically focused efforts to assess the use of M&S and analysis in support of nonkinetic aspects of COIN and IW in the early stages of OEF and OIF. Publication of U.S. Army Field Manual 3-24 (FM 3-24) helped to address this imbalance,[15] but it did not devote significant space to the use of M&S and analysis in support of commanders' decisionmaking. Further, through at least the late 2000s, neither service nor joint doctrine and literature helped to place COIN within the broader context of IW or adequately define IW in order to help direct M&S/analytic efforts. In 2006, an overarching concept of IW was considered "too immature" and "did not have doctrinal utility."[16]

As both OIF and OEF evolved, commanders and analysts were increasingly exposed to COIN and IW literature. In turn, this literature appears to have informed decisions and analysis—commanders routinely reference this literature in postdeployment professional articles in *Military Review* and similar publications. References in military publications to traditional IW "canon," such as works by C. E. Callwell, David Galula, and Roger Trinquier, spiked in 2006 and have remained at a constant level (Figure 2.2).[17] Required reading lists for deploying warfighters began to include traditional COIN literature as well as newly influential works that discussed IW in the modern context, such as those by John Nagl, Thomas X. Hammes, and David Kilcullen.[18] Taken together, the canonical literature and modern publications provided considerable insight into the kinds of questions commanders might need to address in IW, and specifically in COIN.

The shift in operational focus from mostly kinetic, enemy-centric operations to operations balancing offensive operations with civil and information considerations challenged commanders and warfighters with a new set of decision issues. As one commander noted:

---

[15] Headquarters, U.S. Army, *Counterinsurgency*, Field Manual 3-24/Marine Corps Warfare Publication 3-33.5, Washington, D.C., December 2006d.

[16] Joint Warfighting Center, "Irregular Warfare Special Study," U.S. Joint Forces Command, August 4, 2006b, pp. IV–1

[17] C. E. Callwell, *Small Wars: Their Principles and Practice*, 3rd Edition, Lincoln, Neb.: University of Nebraska Press, 1996 (Original Publication 1896); David Galula, *Counterinsurgency Warfare: Theory and Practice*, Westport, Conn.: Praeger Security International, 1964; Roger Trinquier, *Modern Warfare: A French View of Counterinsurgency*, trans. Daniel Lee, New York: Frederick A. Praeger, 1964.

[18] John A. Nagl, *Counterinsurgency Lessons from Malaya and Vietnam: Learning to Eat Soup with a Knife*, Westport, Conn.: Praeger Publishers, 2002; Thomas X. Hammes, *The Sling and the Stone: On War in the 21st Century*, St. Paul, Minn.: Zenith Press, 2004; David Kilcullen, *The Accidental Guerrilla: Fighting Small Wars in the Midst of a Big One*, New York: Oxford University Press, 2009a.

Figure 2.2
COIN and IW References in Select Military Publications

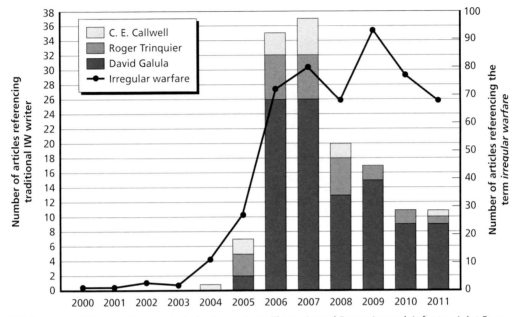

NOTE: Select military publications include: *Army, Army Times, Armed Forces Journal, Infantry, Joint Force Quarterly, Leatherneck, Marine Corps Gazette, Military Intelligence Professional Bulletin, Military Review,* and *National Guard.*

RAND *RR382-2.2*

It certainly did not take long to discover that the traditional tools in my military kit bag were insufficient to successfully compete in this new operational environment. As a brigade commander, I was somewhat surprised to find myself spending 70 percent of my time working and managing my intelligence and IO [information operations] systems and a relatively small amount of my time directly involved with the traditional maneuver and fire support activities.[19]

The vast literature describing personal and unit experiences in OEF and OIF recounts the challenges faced in balancing limited time and resources to achieve long-term strategic goals in varying operational battlespaces. Overall, the literature from this period describes decision issues falling into the following categories:[20]

- Designing the IW strategy
- Coordinating lines of operation (LOOs) with interagency and multinational partners

---

[19] Ralph O. Baker, "The Decisive Weapon: A Brigade Combat Team Commander's Perspective on Information Operations," *Military Review*, May-June 2006.

[20] Based on our interviews, we consolidated this to the four categories presented in Chapters Four through Seven.

- Developing host-nation security and governance capabilities
- Conducting operations to disrupt and defeat enemy combatants/enablers
- Positively influencing the population
- Sustaining the campaign
- Assessing operational performance and effectiveness.

Within each of these broad categories, commanders had to consider the most efficient and effective ways to integrate offensive, defensive, and civil-military information means to achieve tactical to strategic objectives. Emphases on different means and categories varied widely within theaters and over time as each commander grappled with the unique operational environment within which each unit operated. Despite situational differences, commanders and analysts alike encountered similar issues in applying analysis products to the decisionmaking process, which will be discussed in a later section.

M&S and Analytical Methods Identified in Literature

Review of 2001–2012 literature related to OEF and OIF revealed that noncomputational analytic methods, often aided by computational tools such as Microsoft Excel, represented a significant proportion—if not the majority—of analytic support during OEF and OIF. Because technical terms are often and unhelpfully used without distinction in the literature, it is necessary to draw a distinction between M&S, analytic methods, and computational tools as they appear to be represented in existing literature. We have already defined modeling, simulation, and analysis for the purposes of framing this report; this section reflects interpretations of M&S, analysis, computational tools, and what might be called the "analyst decision space" as they apply to IW decision support. By providing a separate set of definitions we intend to show the variation in understanding and approaches; this variation is sometimes helpful but may sometimes be misleading.

Demonstrated IW Decision Support Identified in Literature

The literature reaches a broad consensus that there is a gap in sociocultural analytical capability within DoD. Because IW and, specifically, COIN tend to be population-centric endeavors, this gap has undermined the military's ability to provide adequate analytic support to commanders' decisions. As of 2006, DoD's structured sociocultural analysis derived primarily from academia, national labs, and industry, which focused at the strategic level with limited applicability to commanders at operational echelons.[21] The Defense Science Board's 2007 report, "Understanding Human Dynamics," echoed the community's concerns that the "military belatedly adapted to the human dynamic needs" of the current COIN and IW conflicts, and it identified a multitude of general gaps in data gathering, personnel training, and analytic tools.[22]

By the 2008–2009 time frame, combatant commands and the services began formally investing resources into sociocultural research and analysis, and the emergence of human terrain teams and DoD-internal research began to fill in data gaps. However, despite some uneven progress toward these objectives—most notably in the realm of intelligence analysis—analytic solutions for sociocultural decision issues at the operational and tactical levels are viewed as "long-term goals."[23]

Courses of Action Associated with Families of Analysis

When commanders make decisions in COIN, these decisions tend to result in a mix of kinetic and nonkinetic actions; there was little recorded evidence of commanders deciding to take no action to support a key decision (e.g., deciding to let a situation develop rather than act). A kinetic course of action (COA) entails the employment of individual weapons systems, force maneuver, fires, and the analysis of weapon effects/forensics. A nonkinetic COA, however, includes those activities that do not utilize weapon systems to achieve operational results. These include traditional military COAs such as ISR; engineering operations; and logistics and sustainment. In the IW context, however, there is an increased emphasis on sociocultural-based COAs such as IO and civil operations.

In COIN and IW, the literature focuses on what might be termed two "families" of analysis that can be used to support decisions and to recommend some mix of kinetic or nonkinetic actions. The first family, physics-based analyses, depends primarily on mathematical analysis of easily quantifiable mission variables, such as terrain features, weapon capabilities, and numbers of personnel and equipment. Social science analysis, on the other hand, can entail both mathematical methods and variable values derived through more qualitative methods. For example, social science analysis may include statistical analysis of surveys, cultural analysis performed by SMEs, and organizational dynamics analysis. By presenting the COAs and families of analyses along two axes in Figure 2.3, we can begin to categorize some examples of demonstrated decision support by M&S and analytical methods.

---

[21] Dylan Shmorrow, "Sociocultural Behavior Research and Engineering in the Department of Defense Context," Office of the Secretary of Defense, Assistant Secretary of Defense for Research and Engineering Human Performance, Training, and BioSystems Directorate, August 2011, p. 20.

[22] Defense Science Board, *Understanding Human Dynamics*, Washington, D.C.: Office of the Under Secretary for Acquisition, Technology, and Logistics, March 2007, p. 4.

[23] Shmorrow, 2011, p. 20. See also Defense Science Board, *Counterinsurgency (COIN) Intelligence, Surveillance, and Reconnaissance (ISR) Operations*, Washington, D.C., Office of the Under Secretary of Defense for Acquisition, Technology, and Logistics, February 2011.

Figure 2.3
Interaction of COAs and Analysis Families

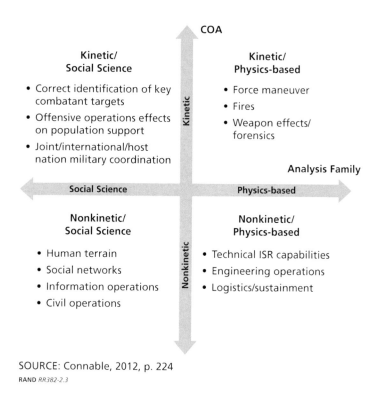

SOURCE: Connable, 2012, p. 224
RAND *RR382-2.3*

Kinetic Activity Decisions Supported by Physics-Based Analysis

Modeling and analysis techniques have an established place in supporting military decisions that tend to result in kinetic activities, particularly those found in the upper-right quadrant of Figure 2.3. The "OR toolkit," or set of methods and tools typically applied by ORSAs, seem to be particularly useful to support this kind of activity.[24] Despite an emphasis on sociocultural aspects in COIN and IW in the literature, physics-based analysis also plays an important part in helping commanders effectively execute the offensive operations necessary to disrupt and defeat enemy combatants. During the clearing stage of the strategy, "Clear-Hold-Build," commanders consider and tend to focus on Blue force maneuvers, fires, and weapons effects/forensics. Commanders must also consider enemy COAs, including their offensive capabilities and tactics. Overall, analysts had a variety of analytic methods, models, and computational tools available

---

[24] Operations research gained considerable influence during the Cold War period, advancing models and simulations covering conventional force-on-force interactions and force capabilities at the strategic and tactical levels. For an overview of operations research, focusing mainly on the Army, see Seth Bonder, "Army Operations Research—Historical Perspectives and Lessons Learned," *Operations Research*, January-February 2002.

to support decisionmaking likely to result in physics-based, kinetic activities, and most analysts had the required expertise and were comfortable using these techniques.

In terms of Blue forces, commanders focused primarily on assessing physical force protection measures and tracking the maneuver of their units. The threat of improvised explosive devices (IEDs), in particular, presented a significant challenge to commanders, and extraordinary fiscal and personnel resources were dedicated to addressing the IED issue. To a great extent, the creation of the Joint IED Defeat Organization (JIEDDO), with billions of dollars in annual funding, hinged on the need to develop specific analytic data and techniques for C-IED and to protect Blue forces. Unfamiliar threats and operational environments prompted the introduction of new technological capabilities into the field, and analysts provided assessment of the technologies' effectiveness, primarily through reachback support. JIEDDO and non-JIEDDO agencies funded, supported, and coordinated a wide array of analyses in support of IW force protection. The Army Material Systems Analysis Agency and the Army Test and Evaluation Command conducted platform performance parameter and weapon system modeling for friendly as well as enemy weapon systems.[25] Chapter Three records several examples of C-IED decisions supported by both analyses and modeling and simulation.

To assist commanders in planning kinetic COAs in the field, analysts relied on commonly used tools and methods to conduct geospatial and terrain analysis for troop movements, using programs such as FalconView and ArcGIS. Some analysts also attempted to use more sophisticated modeling tools, such as 3-D modeling software, to help their unit "walk through" travel routes or target buildings, although access to the necessary computational tools was limited once in theater.[26] The tracking of Blue forces relied upon a variety of computational tools, such as the Maneuver Control System, Force XXI Battle Command Brigade and Below System, the Excel-based Company Movement and Assessment Tool, and Blue Force Tracker (BFT).

Decision support analysis of enemy kinetic actions focused primarily on types and geographical location of attacks. Analysts primarily drew upon Significant Activity data (SIGACTs) recorded within the Combined Information Network Data Exchange to identify geographical and methodological patterns in IED usage, snipers, direct and indirect mortar attacks, and force-on-force engagements with enemy combatants. The data from relevant SIGACTs was typically manipulated to create charts, graphs, and map overlays, or it was analyzed using a statistical package or program. Analysts also examined mortar and IED attacks using traditional crater and weapons-forensic

---

[25] Center for Army Lessons Learned, "ORSA Handbook for the Senior Commander," March 1, 2008.

[26] A computer program, Sextant, was used frequently during several units' training and considered extremely helpful. However, there was no discussion identified of a unit utilizing Sextant during deployment in OIF or OEF. John P. Piedmont, "Det One, U.S. Marine Corps U.S. Special Operations Command Detachment, 2003–2006: U.S. Marines in the Global War on Terrorism," History Division, U.S. Marine Corps, 2010, p. 41.

techniques and pattern analysis to determine most likely time and locations for future attacks.[27]

Pattern and trend analysis constituted the majority of analytic methods utilized, occasionally bolstered by statistical methods and computational tools. Typical computational tools used in analyzing enemy kinetics were Microsoft Excel (including advanced package add-ins), geospatial tools such as ArcGIS and FalconView, and various statistical programs. Advanced computational tools, however, were not always available in theater, and analysts often had to resort to more manual means of analysis, such as using "Microsft Excel spreadsheets and grease pencils" to track enemy activities.[28]

Despite established expertise and tools to conduct physics-based analysis of kinetic activity in the COIN and IW context, this form of analytic decision support in isolation could not provide a comprehensive picture for the commander. Targeting, especially of key individuals within an insurgent group, requires a greater reliance on human intelligence and link and social network analyses. Traditional maneuver force-on-force methodologies fell short in "man hunting" critical nodes in adversary organization. The data used to study friendly and enemy force maneuvers and tactics proved to be largely incomplete, inaccurate, or difficult to manipulate or analyze in a meaningful way to the warfighter. Also, as shown before, the objective of destroying enemy targets is one of many interconnected LOOs within COIN and IW, and a "successful elimination" may have adverse effects upon the other COIN and IW objectives.

Kinetic Activity Decisions Supported by Social Science Analysis

Although most of the literature on COIN and IW modeling and analysis equates social science analysis primarily with nonkinetic activities, when commanders make decisions that lead to kinetic actions these actions can impact the socio-cultural environment. As one observer noted, "the holy grail of irregular warfare simulation is modeling second- and third-order effects."[29] When making decisions that lead to the execution of violence, commanders must also consider the population's reaction to the target's removal and the method used, the effects on future human intelligence human intelligence collection, and the kinetic operation's effect on the local power balance. As General Stanley McChrystal observed, the tendency of the military to focus primarily on direct kinetic effects "masks the true extent of the insurgent activity and prevents

---

[27] See Edward J. Coleman and Rico R. Bussey, "A Primer on Indirect Fire Crater Analysis in Iraq and Afghanistan," *Field Artillery Journal*, July-August 2005; T. J. Ramjeet, "Operational Analysis in Support of HQ RC(S), Kandahar, Afghanistan, September 2007 to January 2008," in *The Cornwallis Group XIII, Analysis in Support of Policy*, Farnborough, UK: Defence Science and Technology Laboratory, 2008.

[28] Trey Birdwell and John A. Klemunes, "Tools of War," *Engineer*, January-March 2004, p. 36.

[29] Michael Peck, "Firmer Ground," *Training and Simulation*, Vol. 14, October 1, 2011.

an accurate assessment of the insurgents' intentions, progress, and level of control of the population."[30]

Early in both OEF and OIF, commanders and analysts "reverted to counting specific numbers of targets destroyed to determine combat progress" and "fell back to assessing what they knew or could assess in the allotted time, mostly tried-and-true measures of attrition."[31] Official doctrine suggests utilizing opinion polls and surveys to gauge second- and third-order effects, though in practice these methods have proved subject to method and sampling errors due to lack of safe access to populations and limited resources and expertise.[32] Comprehensive, long-term polling efforts outside DoD, such as the annual "Survey of the Afghan People" conducted by the Asia Foundation, offer validated insight, though these do not directly link certain kinetic actions to population reactions.[33] Analysts have also relied on proxy measures—such as number of IED placements reported by civilians, numbers of civilian casualties, or local media content analysis—to conduct limited trend and pattern analysis.[34]

One attempt to systematically analyze population reactions to Blue force actions is the Joint Non-Kinetic Effects Model (JNEM) developed by the Army. The mission-readiness exercise in preparation for OIF deployment, Unified Endeavor, used JNEM software beginning in 2006. Though "deemed operationally successful," JNEM has many theoretical and interface issues that hamper its applicability to OEF and OIF environments.[35] PSOM, addressed in more detail in the appendix, also attempts to analyze the nonkinetic effects of kinetic COAs, as well as nonkinetic COAs. Created to fill the need for an analytical tool outside those for unit-to-unit combat, PSOM has been used by the J8 Warfighting Analysis Division and coalition partners to help produce and inform campaign planning.[36] Overall, most M&S tools applying social science to plan and assess kinetic actions were used primarily at the strategic level or as predeployment training aids.

---

[30] DoD, "COMISAF Initial Assessment (Unclassified)," Republished in *The Washington Post*, September 21, 2009c.

[31] Draft JFCOM lessons learned report quoted in Elaine M. Grossman, "JFCOM Draft Report Finds U.S. Forces Reverted to Attrition in Iraq," *Inside the Pentagon*, March 25, 2004.

[32] For further discussion of these pitfalls see: Todd C. Helmus, Christopher Paul, and Russell W. Glenn, *Enlisting Madison Avenue: The Marketing Approach to Earning Popular Support in Theaters of Operation*, Santa Monica, Calif.: RAND Corporation, MG-607-JFCOM, 2007, pp. 47–48.

[33] Asia Foundation, *Afghanistan in 2011: A Survey of the Afghan People* web page, undated.

[34] See David F. Eisler, "Counter-IED Strategy in Modern War," *Military Review*, January-February 2012.

[35] For a more thorough discussion of this model, see Hugh Henry, "A Non-Kinetic Effects Federate for Training Simulations," *Journal of Defense Modeling and Simulation: Applications, Methodology, Technology*, Vol. 6, No. 3, 2009.

[36] For a more thorough discussion of this model, see Howard Body and Colin Marston, "The Peace Support Operations Model: Origins, Development, Philosophy and Use," *Journal of Defense Modeling and Simulation: Applications, Methodology, Technology*, Vol. 8, No. 2, 2010.

In the literature, analysts describe using techniques ranging from "using pads of butcher-board paper, yellow stickies, and a large wall chart" to using computational tools such as IBM's Analyst's Notebook.[37] Link and influence diagrams appeared to help commanders visualize enemy networks and outside connections, but descriptions of how one prioritized targets using analytic methods were sparse.

In both OEF and OIF, success depended in large measure on the development of sufficient and competent host-nation security forces. Most commanders described an interest in assessing these forces, even when it was not considered their primary responsibility (e.g., a tactical commander as opposed to a mentor or training officer). To meet this requirement, assessment tended to focus on trends in the numbers of trained personnel, equipment, vehicles, and units reaching certain MOEs.[38] Assessment processes in and across OIF and OEF varied, but analysts in both theaters faced similar challenges in collecting useful data. Analytic techniques used to assess security force assistance (SFA) effectiveness, such as the Commander's Unit Assessment Tool in Afghanistan and Operational Readiness Assessments in Iraq, were subject to criticism.[39] As one former combat adviser observed, "This data is a snapshot of health in an organization. What these do not tell the adviser is cause . . . there are also some areas of professional growth and maturity in a host-nation unit that simply cannot be measured."[40] Outside of assessment tools, several efforts to aid commanders in security force assistance decisions included CAA's Afghan National Security Force (ANSF) generation simulation model, and a discrete event simulation called the ANSF Growth and Retention Analysis Model. Another example of analytic support for security force assistance was the Marine Corps Combat Development Command's regression analysis of Afghanistan data derived from the Commander's Unit Assessment Tool.[41]

Nonkinetic Activity Decisions Supported by Physics-Based Analysis

There were few mentions in the open literature of analytic challenges associated with COIN and IW tasks such as the allocation of ISR assets or the prioritization of engineering activities. Analyses of pathway risk for logistics and sustainment, however,

---

[37] Lester W. Grau, "Something Old, Something New: Guerrillas, Terrorists, and Intelligence Analysis," *Military Review*, July-August 2004.

[38] For a high-level example of trend assessment, see the annual DoD reports titled *Toward Security and Stability in Afghanistan* and *Measuring Stability and Security in Iraq* on DoD's website.

[39] For examples, see Office of the Special Inspector General for Afghanistan Reconstruction, *Actions Needed to Improve the Reliability of Security Force Assessments*, SIGAR Audit 10-11, June 29, 2010; Office of the Special Inspector General for Iraq Reconstruction, *Iraqi Security Forces: Police Training Program Developed Sizeable Force, but Capabilities Are Unknown*, SIGIR 11-003, October 25, 2010.

[40] Thomas Seagrist, "Combat Advising in Iraq: Getting Your Advice Accepted," *Military Review*, May-June 2010, p. 71.

[41] David Smith, "CAA Current Operations Support to OIF/OEF," Presented at the 49th Army Operations Research Symposium, October 13–14, 2010.

appear to have presented considerable difficulty. Most analysts found themselves confined to simply plotting attack density and timing along key routes. Logisticians utilized database spreadsheets and geospatial tools to plot IED and small arms attacks along critical supply routes, although these methods were "not infallible," and entailed a heavy reliance on ISR products.[42] Logistics databases typically consisted of "an intricate system of spreadsheets" which proved to be fragile and labor intensive."[43] One tool for helping to manage all the data in real time is the Logistics Reporting Tool, a component of the Battle Command Sustainment Support System spearheaded by the First Infantry Division operations staff in Iraq. The tool, however, has had mixed reviews because of its reportedly unwieldy and unintuitive interface.[44]

Analysts also supported commanders' decisions concerning line of sight and system placement for ISR components. A common analytic technique was to utilize geospatial and trend analysis of enemy IED and small arms attacks to more effectively place ground-based and aerial ISR. In particular, analysts used crime-mapping techniques to develop temporal and spatial algorithms to identify attack characteristic patterns for ground-based systems placement.[45] The mass amounts of sensor data, however, overwhelmed processing, exploitation, and dissemination capabilities of the analysts and often prevented them from engaging in higher-level analysis on the population and insurgent group behavior.[46]

Nonkinetic Activity Decisions Supported by Social Science Analysis

Throughout the OEF and OIF literature, firsthand accounts abound concerning the military's limits and ad hoc approach in both social science analysis and sociocultural-based nonkinetic activities, particularly in sociocultural situational awareness and IO. Analysts struggled to aid commanders' decisions concerning the prioritization, execution, and assessment of civil operations and IO, especially within the context of continuing offensive operations within respective areas of responsibility (AoRs).

Commanders and analysts alike relied upon a "dizzying array of acronyms and terms" to help them compartmentalize and understand the complex sociocultural

---

[42] Mary K. Kahler, "Providing S-2 Support for a Brigade Support Battalion," *Army Logistician: Professional Bulletin of United States Army Logistics*, November-December 2008.

[43] Larry L. Motley, "Developing a Fuel Management Information System in Iraq," *Army Sustainment*, September-October 2011, p. 48.

[44] Benjamin Kibbey, "1st Infantry Division Recognizes Benefits of Logistics Reporting Tool," *Army Sustainment*, November-December 2010; Michael B. Siegl, "Sustaining a BCT in Southern Iraq," *Army Sustainment*, November-December 2010.

[45] See James A. Russell, "Innovation in War: Counterinsurgency Operations in Anbar and Ninewa Provinces, Iraq, 2005–2007," *Journal of Strategic Studies*, Vol. 33, No. 4, 2012; Scott Kinner, "Expanding Attack the Network," *Air Land Sea Bulletin*, September 2012, pp. 4–7.

[46] Defense Science Board, 2011.

environments of Afghanistan and Iraq.[47] The most frequent acronyms used to help dissect the sociocultural environment included DIME (Diplomatic, Informational, Military, Economic), ASCOPE (Areas, Structures, Capabilities, Organizations, People, and Events), and PMESII-PT (Political, Military, Economic, Social, Information, Infrastructure, Physical Environment, Time).[48] The listing of these variables may have provided a more digestible picture of the complex environment, but the relationships between each proved more difficult to understand.[49] Summing up a general consensus of the literature, one commander notes:

> Military professionals describe this volatile mix of factors as being ambiguous, complex, uncertain, and ill-structured. When trouble appears, there is no consensus about what the fundamental problems are, how to solve them, what the desired "end state" should be, and whether an "end state" is achievable or not.[50]

While it was possible to envision an end state for insurgents—kill, capture, or coopt—it was more difficult to envision end states that incorporated broader sociocultural factors. Commanders struggled to find ways to communicate with populations in both Iraq and Afghanistan. Many commanders interviewed or who wrote articles or books emphasized the need to improve their IO, though expertise and analytic tools were also often lacking.[51] Doctrinal guidance on COIN and IW also stresses the importance of IO as "setting the conditions" for the success of all other COAs taken.[52] At least during the early phases of OEF and OIF, IO responsibilities were often delegated to those with limited to no experience in the field, and IO typically only became a focus after initial offensive, clearing operations had been completed. At least through the late 2000s, lessons learned documents, service, and joint-level manuals offered little guidance or assistance in designing proper analytic techniques to plan and assess IO integration into the commander's other LOOs.[53] Commanders and analysts

---

[47] Celestino Perez, "A Practical Guide to Design: A Way to Think About It, and a Way to Do It," *Military Review*, Vol. 91, No. 2, March-April 2011.

[48] For a full description of each of these constructs, see Jack D. Kem, "Understanding the Operational Environment: The Expansion of DIME," *Military Intelligence Professional Bulletin*, Vol. 33, No. 2, April-June 2007.

[49] An example of an analytic framework utilizing these constructs is Peter R. Mansoor, "Linking Doctrine to Action: A New COIN Center-of-Gravity Analysis," *Military Review*, September-October 2007.

[50] Perez, 2012.

[51] For examples of IO analysis and execution, see Baker, 2006; Thomas F. Metz, Mark W. Garrett, James E. Hutton, and Timothy W. Bush, "Massing Effects in the Information Domain: A Case Study in Aggressive Information Operations," *Military Review*, May-June 2006; Joseph F. Paschall, "IO for JOE: Applying Strategic IO at the Tactical Level," *FA Journal*, July-August 2005.

[52] Headquarters, U.S. Army, 2006d, pp. 5–8.

[53] See Arturo Munoz, *U.S. Military Information Operations in Afghanistan: Effectiveness of Psychological Operations 2001–2010*, Santa Monica, Calif.: RAND Corporation, MG-1060-MCIA, 2012.

also asserted that translating IO objectives from the strategic to the tactical level was difficult, since strategic messages may receive different receptions depending upon the tactical AoR's cultural makeup.[54] One example of disparity between different echelons involves the importance of analyzing enemy IO themes and delivery platforms. In a recent RAND study, battalion- and brigade-level commanders did not place significant importance on collecting and analyzing enemy IO activity, but this activity did occur at higher levels.[55]

Despite the lack of capabilities and higher-level guidance, many units adapted analytic capabilities to perform the necessary target audience analyses required for their unique operational environment's IO. For example, a commander would assess IO effectiveness by performing surveys, local media content analysis, or "reality checks" with experts or locals on the social and cultural implications of the message. Other measures of effectiveness gathered to perform pattern and trend analyses included numbers of local tips on insurgent activity, anti-U.S. graffiti, and mosque sermons' content. However, literature reviewed for this report gave the impression that lower-level commanders assessed IO effectiveness through the response of local power brokers rather than engaging in surveys or studying behavioral trends. Pattern and trend analysis aided by computational spreadsheet and statistical tools constituted a majority of analysis performed for operational and tactical level IO. Strategic-level analysis occasionally relied upon M&S tools, such as the PSOM or Senturion, to help commanders understand the ways in which IO might affect a targeted population.[56]

Another important decision commanders faced concerning nonkinetic activities was the coordination of COAs between various actors, including the State Department, nongovernmental organizations, and international partner organizations.[57] The survey of the literature, however, uncovered very little analysis that aided decision support on this issue.

Pattern and Trend Analysis and Assessment

Contemporary literature on COIN and IW tends to suggest that pattern and trend analysis serve as the default approach to understanding the complex environment and to explain progress in a campaign. Pattern analysis is the identification of similarities in activity either over time or in one area all at once. For example, ISAF identi-

---

[54] As discussed in Center for Army Lessons Learned, "Operation Iraqi Freedom: Information Operations, Civil Military Operations, Engineer, Combat Service Support," Initial Impression Report No. 04-13, May 2004a.

[55] Eric Larson, Richard E. Darilek, Dalia Dassa Kaye, Forrest E. Morgan, Brian Nichiporuk, Diana Dunham-Scott, Cathryn Quantic Thurston, and Kristin J. Leuschner, *Understanding Commanders' Information Needs for Influence Operations*, Santa Monica, Calif.: RAND Corporation, MG-656-A, 2009b, pp. 8–9.

[56] See Mark Abdollahian et al., "Senturion: A Predictive Political Simulation Model," Center for Technology and National Security Policy, National Defense University, July 2006.

[57] For a discussion on this difficulty, see Lewis G. Irwin, "Irregular Warfare Lessons Learned: Reforming the Afghan National Police," *Joint Force Quarterly*, No. 52, 1st Quarter, 2009.

fied a correlation between seasonal weather patterns and the frequency of violent incidents year-on-year from 2004 to 2010, showing that attacks tended to decrease during the winter and increase during the summer.[58] Trend analyses portray "observed" changes in behavior in a specific variable (e.g., attack incidents) over a specific period of time and in a designated space (e.g., all of Afghanistan, or perhaps a single district). FM 3-24 is particularly aggressive in recommending the use of pattern and trend analysis, and provides archetypal charts and tables.[59] Exemplifying the general view within the literature, one Company commander asserted:

> Bottom line—the only way to get ahead of the enemy's decision cycle is to constantly and thoroughly analyze every scrap of information you can get your hands on and try to "see" patterns.[60]

But these kinds of analyses often constituted little more than "collation and presentation of data rather than extensive analysis."[61] For example, one analyst describes the reaction of one commander to a trend analysis presentation: "Attacks fell below the twelve week average for the first time in 2008. Whoop de doo! You OA types need to provide me with some analysis beyond bar charts."[62]

Without a structured, repeatable analytic approach to choosing the variables to track, identifying *significant* trends, and assessing causal relationships, the "eyeballing" approach could possibly lead to misconstrued conclusions. Aware of this shortfall, one analyst posited:

> . . . statistical analysis is the only means that allow determination of a change (effect) with a specified degree of confidence that the change is real, and not just due to random chance. In some cases, apparent effects may be so numerically large that they appear to obviously represent true changes. However, even apparently large effects may actually be due to random chance. There is no substitute for proper statistical analysis.[63]

---

[58] Eric Gons, Jonathan Schroden, Ryan McAlinden, Marcus Gaul, and Bret Van Poppel, "Challenges of Measuring Progress in Afghanistan Using Violence Trends: The Effects of Aggregation, Military Operations, Seasonality, Weather, and Other Causal Factors," *Defense and Security Analysis*, Vol. 28, No. 2, June 2012, p. 102.

[59] Headquarters, U.S. Army, 2006d.

[60] John Paganini, *Counterinsurgency Lessons Learned*, DoD, November 16, 2011, pp. 18–19.

[61] E.g., see D. J. Evans, "Operational Analysis in Support of HQ ISAF, Kabul Afghanistan, 2002," in *The Cornwallis Group VIII: Analysis for Governance and Stability*, Farnborough, UK: Defence Science and Technology Laboratory, 2002. Also see Connable, 2012.

[62] Robert Shearer, "Operations Analysis in Iraq: Sifting Through the Fog of War," *Military Operations Research*, Vol. 16, No. 2, 2011.

[63] Anthony E. Pusateri, "Metrics to Monitor Governance and Reconstruction in Afghanistan: Development of Measures of Effectiveness for Civil-Military Operations and a Standardized Tool to Monitor Governance Qual-

However, as the analyst later notes, the expertise and computational tools required to perform such analysis may not be readily available in theater and should therefore be delegated to reachback support. Whether these analyses were conducted in theater or not, a range of Vietnam-era literature as well as some more recent information identified concerns with the ability to conduct sufficiently accurate pattern and trend analyses, and also with these analyses' relevance to decision support. Gons et al. identify some of the dangers inherent in aggregating quantitative data for pattern and trend analysis:

> [I]t is important to keep in mind the consequences and potential limitations inherent in aggregate analyses. At face value, an aggregated analysis allows a reader to glean a top-level understanding of subject matter while not getting mired in every detail. In the cases where data have widely differing magnitudes or exhibit disparate trends, the resulting aggregate analysis can be incomplete or misleading. This is particularly true in complex domains where the factors that impact the analysis are numerous, may not be well-understood, or cannot easily be incorporated into a model. Such is the case in Afghanistan, with numerous physical, geographic, climatic, and social/ideological boundaries.[64]

Further, it is not clear that patterns or trends constitute sufficient support to decisions. Stephen Downes-Martin of the U.S. Naval War College points out that, "In the absence of a credible numbers-based theory of counterinsurgency there can be no objective, numbers-based assessment for Operation Enduring Freedom."[65] Therefore, he argues, the quantitative approaches delineated in current U.S. and North Atlantic Treaty Organization (NATO) doctrine are not appropriate for at least campaign assessment and strategic assessment.

Pattern and trend analyses sit astride the dividing line between analysis and assessment. These analytic efforts tend to compare large data sets using thoroughly validated and commonly applied statistical or other analytic techniques. There may be nothing wrong with the techniques themselves. Even when the data available for pattern and trend analyses are inaccurate and incomplete, it is still possible to deliver meaningful results if the variables are chosen carefully and the analysts are careful to explain the limitations of their tools and methods. Some of these quantitative analyses were of high quality while others were not; often the data are poor and incomplete, yet they are presented as accurate and complete; sometimes analysts explained the value

---

ity," U.S. Army Civil Affairs and Psychological Operations Command, Technical Report 04-01, March 12, 2004.

[64] Gons, Schroden, McAlinden, Gaul and van Poppel, 2012, p. 101. For a detailed analysis of these issues, see Connable, 2012.

[65] Stephen Downes-Martin, "Operations Assessment in Afghanistan Is Broken: What Is to Be Done?" *Naval War College Review*, Newport, RI: Naval War College Press, Vol. 64, No. 4, 2011, p. 103.

of their reports clearly and effectively but in other cases failed to successfully interpret their own work. Yet the *idea* of pattern and trend analysis is not necessarily anathema to a better understanding of COIN and IW.

Assuming a competent analyst presents a carefully bounded product with appropriate caveats, the most pressing concern may be with the way these analyses are read by commanders and incorporated into both assessments and decisions. Commanders read quantitative pattern and trend analyses and then must decide what they mean and whether they are meaningful enough to influence decision. Sometimes pattern and trend analyses show results that are valid but meaningless in terms of campaign progress, while in other cases they can be misleading in their convincing precision. In still other cases, commanders eager for clarity in the complex world of COIN and IW read too much into these analyses. In this latter and seemingly common case, the fault lies not with the method, the analyst, or the data, but with the use of the analytic product. Some commanders are considered good consumers—they have a background in analysis and understand the vagaries of the methods and data. Others are unprepared to employ ORSAs or other analysts properly, and are similarly unprepared to read their work with a sufficiently critical eye. Successful use of pattern and trend analysis in COIN and IW requires both a competent analyst and an educated, thoughtful commander.

Literature Identifies Factors That Hindered Analytic Support to Commanders

Analysts encountered several factors that hindered analytic support to commanders outside of available computational tools and analytic methods. Of these, data management and relevancy of analysis to the warfighter were the most discussed within the literature. Commanders and warfighters faced daily battles in balancing kinetic and nonkinetic activities to achieve often nebulous objectives, but, as one analyst posits, ". . . behind COIN outside the wire is COIN as practiced in spreadsheets, slide presentations, and link charts on the FOBs [forward operating bases] across Iraq and Afghanistan."[66] The demands for quick turnaround analysis and recurring analytic product requirements often forced analysts to "automate" analysis through Excel pivot tables and PowerPoint templates. However, with regular turnover and frequent transfer of data between units and echelons, "juggling scores of Microsoft Excel spreadsheets or homegrown Access databases is a recipe for information management failure."[67] In addition to fragile data tools, analysts also had to contend with a multitude of stovepiped data repositories of varying quality, completeness, and access control. The Iraq theater at one time had more than 300 different databases, and "even within the same

---

[66] J. Edward Conway, "Analysis in Combat: The Deployed Threat Finance Analyst," *Small Wars Journal*, July 5, 2012.

[67] Morgan G. Mann, "Thoughts Regarding the Company-Level Intelligence Cell," *Marine Corps Gazette*, June 2009.

warfighting function—logistics, for example—all users could not see the data."[68] Finding databases in Afghanistan had become so difficult by 2010 that one analyst started a program called "data cards," a project designed to identify sources of data through crowdsourcing rather than relying on what had evolved into a failed systemic approach to data-sharing.[69]

Many analysts also faced difficulty in proving and then maintaining relevancy to the warfighter in terms of both their analysis subject matter and the presentation of their conclusions. Some of the firsthand accounts from analysts in the field recount the difficulty some commanders had in incorporating analyst products into their decision-making process, because of either their individual predilections or their unfamiliarity with the capabilities of the analysts in their units. ORSAs deployed to OIF and OEF seemed to encounter great hurdles in proving their worth to commander decision support. One analyst related his own experience with 16 Army majors:

> I received quite a few perplexed looks after introducing myself . . . the first question was, "What the hell is an ORSA? I have never heard of that before." My answer was, "Well . . . we do math and stuff to help the commander make better decisions." As I said math, the eyes of my cohorts rolled in their heads, and the focus shifted to another subject.[70]

As the need for analysis increased, particularly in support of what many considered to be unorthodox, nonkinetic IW COAs, analysts sought to create relevant products for their commanders. Geospatial analysis tools, such as FalconView and ArcGIS, helped them to visualize data in a way that aided a commander's thought process concerning the operational battlespace. As one analyst observed during his time supporting command decisions, "Maps are the medium by which they communicate, not plots, charts, nor graphs."[71] Since the beginning of OIF and OEF, however, analysts have become more familiar with a range of tools and display products, and have become more efficient in displaying information. There have also been some efforts to educate commanders to assist them in getting the most out of their analytic support. Agencies and groups such as CAA and NATO's Research and Technology Organisation have sought to educate commanders through "Analyst Handbook" publications, giving an overview of analyst capabilities and where one may incorporate their prod-

---

[68] John R. Vines, "The XVIII Airborne Corps on the Ground in Iraq," *Military Review*, September-October 2008.

[69] This program became so popular that it was expanded to a worldwide effort. See the DataCards homepage, undated.

[70] Sam Sok, "ORSA: Operations Research Systems Analysts Help Develop Solutions," *Infantry*, September-October, 2011.

[71] Shearer, 2011, p. 64.

ucts.[72] However, commanders' approaches to analyses tend to remain idiosyncratic, and to some extent will always be so.

Analysts had to contend with not only data and tool availability along with commander preferences, but also constraints on time and personnel. One analyst concluded that analytic techniques "used in-theater differ greatly from the techniques seen in textbooks and academia, but it is important to remember . . . that time is a luxury the deployed OA [ORSA] teams seldom have."[73] Limited time and resources therefore often made analytic innovation difficult.

Operations Researchers Have Their Say About IW Assessment: MORS Workshops

In 2010, and again in 2012, the Military Operations Research Society (MORS) held workshops designed to identify shortcomings in IW (specifically COIN) assessment, and to recommend ways of improving assessments.[74] In this case, assessment was taken to encompass most of the analytic tasks ORSAs might undertake in IW, but it focused to a great extent on campaign assessment. While many such working groups and conferences have been held between 2009 and 2012 in both the United States and allied countries (particularly The Netherlands), the MORS workshops have brought together most of the relevant experts in the assessment field—military and civilian—and have focused on the most pressing issues that U.S. IW analysts have faced since 2001. This section touches on some of the most applicable findings from these workshops. It is important to note that these reports are necessarily focused on the use of operations research and the role of ORSAs in the assessment process. Findings derive from the assumption that OR and tools from the OR toolkit (e.g., statistical packages, operational modeling) are the appropriate process for IW assessment. This is an assumption that has essentially gone unchallenged at DoD and NATO.

The 2010 workshop included a working group on campaign assessment that was led by the former director of ISAF's Afghan Assessment Group (AAG). This group provided a description of assessment along with objectives for assessments in theater. The definitions of assessment and analysis were disputed even within this workshop.

> Aligned and integrated assessments within a theater must speak with one voice, establish manageable expectations, and serve many masters. First and foremost, assessments must serve the commander and support decision-making. Assessments must add value "down and in" to the subordinate commanders whose assets are being tasked to collect and analyze. Assessments must support reporting requirements "up and out," often through multiple chains of command to national gov-

---

[72] See CAA, 2008; North Atlantic Treaty Organisation, "Decision Support to the Combined Joint Task Force and Component Commanders," Report prepared by the Research and Technology Organisation, Analysis and Simulation Panel, TR-SAS-044, December 2004.

[73] Ramjeet, 2008.

[74] Results from the 2012 MORS workshop were pending release as this report was undergoing editing.

ernments in order to satisfy national security objectives. The proper alignment and integration of assessments is critical to achieve one voice and minimize competition among messages and across resources.[75]

The first working group's report goes on to suggest that, at least through 2010, these objectives were not being met and the lack of adequate assessment doctrine undermined efforts to support commanders' key decisions:

> Despite the critical role that assessments play, organizations frequently treat assessments as an afterthought. Assessment capabilities are often recognized as lacking well after deployment and are subsequently generated out of the institutional force as a temporary loan. A lack of "operating force" assessment doctrine and analytic structure at echelons above corps may contribute to this assessment lag.[76]

This same group identified the lack of adequate metrics, poor understanding of the environment, and excessive flux in analytic methods as sources of poor assessment performance at the operational and strategic level in OIF and OEF. Claflin et al. wrote:

> One of the results of this lack of . . . knowledge is a consistent change to assessment processes. This flux causes a lack of consistent assessments, leading to lack of an ability to determine progress, and also leading to a lack of consistent data over time that can be used to further our understanding.[77]

This follows the assumption that it is necessary to find the "right metrics" to help determine progress in a military campaign, an idea discussed in greater detail in Chapter Five. The second 2010 MORS working group focused on data management for IW analysis. It identified a schism between those analysts who sought greater clarity from commanders in helping them to identify those decisions that required analytic support, and those analysts who were focused on data: "[T]here was . . . a recurring conflict, never resolved [among the working group], between trying to define the questions to be answered and thus driving data requirements, as contrasted with many who wanted data for data's sake . . ."[78] This same group noted that:

---

[75] Bobby Claflin, Dave Sanders, and Greg Boylan, "Improving Analytical Support to the Warfighter: Campaign Assessments, Operational Analysis, and Data Management, Working Group 2: Campaign Assessments," working group briefing, Military Operations Research Society conference, April 19–22, 2010, p. 2.

[76] Claflin, Sanders, and Boylan, 2010, p. 2.

[77] Claflin, Sander, and Boylan, 2010, p. 7.

[78] Michael Baranick, David Knudson, Dave Pendergraft, and Paul Evangelista, "Improving Analytical Support to the Warfighter: Campaign Assessments, Operational Analysis, and Data Management, Working Group 1: Data and Knowledge Management," briefing, Military Operations Research Society, April 19–22, 2010, p. 1.

[T]he current situation is one characterized by stove-piping of information and rampant "ad-hocery" of ORSA analyst organization, institutionalization, and information flow, with little retention of lessons learned or tools and [tactics, techniques, and procedures] developed over the course of U.S. prosecution of [operations in] Iraq and Afghanistan.[79]

A third working group focused on improving analytic support to the warfighter also had trouble identifying the difference between analysis and assessment. This same debate dominated two working groups in the 2012 MORS IW assessment conference, indicating that this is an enduring and unresolved issue that affects the way analysts think about and execute their work. The 2010 group built a pair of working definitions they described as unsatisfying: "There was no formal concurrence among the group. However, to simplify ensuing discussion we agreed that assessment was a focus on key MOE and MOP [measures of performance] that supported lines of effort and analysis were components or building blocks to that assessment."[80] This group identified analyses being conducted in five "key areas," including current operations, planning, future operations, logistics, and intelligence.[81] Most analyses were "lethality focused," but they had a "recognition that nonlethal (governance and economic) [analyses] are important." The report noted that based on the working group's input, 80 percent of analysis in Iraq and Afghanistan was self-initiated rather than commander-driven. "[T]he analyst would listen to the commander, interact with the staff, and follow current and future operations and decide where best to apply themselves."[82] There is no empirical data to substantiate this figure; interviews for this report suggested a varying mix of self-initiated vs. commander-driven work.

The fourth working group focused on OA at the operational and strategic levels of war within IW/COIN. This group identified six distinct categories of analysis that differed from those identified by the fourth group. These included:[83]

1.   **Force structure and force development:** What were the required size and mix of coalition forces and national security [host-nation] force? How effective is the

---

[79] Baranick, Knudson, Pendergraft, and Evangelista, 2010, p. 2.

[80] Thomas Cioppa, Loren Eggen, and Paul Works, "Improving Analytical Support to the Warfighter: Campaign Assessments, Operational Analysis, and Data Management, Working Group 4: Current Ops Analysis—Tactical," briefing, Military Operations Research Society, April 19–22, 2010, p. 3.

[81] Cioppa, Eggen, and Works, 2010, p. 4.

[82] Cioppa, Eggen, and Works, 2010, p. 4.

[83] Greg Graves, Patricia Murphy, and Frederick Cameron, "Improving Analytical Support to the Warfighter: Campaign Assessments, Operations Analysis, and Data Management, Working Group 4: Current Operations Analysis—Strategic and Operational Level," briefing, Military Operations Research Society, April 19–22, 2010, pp. 5–6. Some parts of this list are paraphrased or expounded upon from the original in order to provide clarity.

development of the national security force? How can we most effectively draw down our forces as the conflict subsides?

2. **Size, nature, composition, and objectives of insurgent groups:** Who are the insurgents, what are they capable of, and what are their intentions?

3. **Campaign analysis:** Conduct COA analysis, identify the likely duration of the campaign, determine the probability of campaign success, and help commanders determine cause and effect within the scope of the campaign.

4. **Predicted impact on population of coalition force sizing:** How is the population likely to be impacted by the growth or reduction of coalition forces?

5. **Threats:** Conduct trend analysis of attacks and threats and identify their impact on operations; understand and defeat the enemy network.

6. **Effect of stabilization operations:** How do these operations affect governance, essential services, the security environment, the economy, employment, rule of law, security sector reform, strategic communications, and IO?

The second task—understanding the insurgency—seems to fall clearly within the realm of intelligence analysis. Yet in practice, ORSAs, social scientists, and others have delved into intelligence-like analyses. In some ways this might be helpful: Applying different perspectives and toolsets to such a confounding problem might provide new insight. However, in some cases this could lead to duplication of effort and a reduction in analyst availability for other, perhaps more relevant tasks.

Taking all of the findings from the 2010 workshop together and comparing them with the other literature from 2001 to 2010, it becomes clear that the analytic community did not have a clear understanding of its role in supporting commander's key decisions in IW, at least as of the time of the 2010 workshop—nearly ten years into OEF and seven years into OIF. None of the working groups agreed on a definition of either analysis or assessment, on the specific objectives of analysis in IW/COIN, or on the proper approach to supporting key decisions. However, there was broad consensus reached independently on several points, at least in the 2010 workshop. Most of these conclusions are reflected in our 2012 interviews with ORSAs and commanders, and also in other contemporary literature:

1. Data for analysis and assessment in COIN are often of very poor quality, hard to obtain, and not always relevant to key decisions. Problems with data are compounded when the commander and analyst fail to clarify operational objectives and key decisions—analysts need some semblance of operational clarity to help them focus their efforts.

2. While analysts may initiate a good deal of their own analysis, they must stay focused on supporting the commander's key decisions; analysis and data management must not become ends unto themselves.

3. Analysts must work with not only the commander, but also the military staff and other analysts to ensure they obtain operational context. Analysis done in a vacuum is unlikely to be useful or accepted by IW practitioners.

4. IW/COIN is complex and multifaceted, so analysts must be adaptable. Analysts who can reach into a deep toolkit, think critically, and avoid becoming wedded to a single approach are likely to succeed. "One-trick ponies" are likely to fail to support decisionmaking, and they might unintentionally mislead the people they are trying to support.

Summary of Decision and Analytic Issues in the Literature

Literature on COIN and IW makes it clear that commanders executing these campaigns are faced with a staggering array of complex and interrelated decision points and must act on limited, often inaccurate, and sometimes misleading information. Simply determining which decisions are most important and should be given highest analytic priority can be an exercise in frustration, since the simplest and seemingly least important issue can elevate to the level of strategic importance without warning.

Literature from the OEF and OIF periods revealed that there has been only limited use of complex modeling in both theaters, limited use of simulation, and that most of this support is provided to commanders at the division level and above. Several reports featured the use of modeling, and many analysts built simple models to help them design tools in the field or from reachback support for specific analytic problems. Most modeling consisted of simple back-of-the envelope constructs that helped shape the creation of Microsoft Excel tools. However, intensive modeling efforts that required weeks or months to complete were featured far less often. Since this is a selective literature review and not a comprehensive one, we certainly missed or did not include many examples of modeling in support of IW, particularly if these were noted in obscure publications or contained proprietary information. The selective literature review indicated that the use of analytic tools, specifically Microsoft Office Suite, ArcGIS, and FalconView, was far more commonplace than the use of modeling. Simulations, or wargames, were used in a few cases to support decisionmaking, and we address some of these in Chapter Three.

Analysis in its various forms occurs throughout all levels of command down to the battalion (or tactical) level. In many cases, analysis conducted at the tactical level is rudimentary. Organized analytic support (e.g., the use of ORSAs, social scientists, and other experts) tends to reside at the brigade level and above. While much of this analysis is well considered and of high quality, some of what passes for analysis is simply the presentation of pattern and trend data charts or basic statistical correlation products lacking contextual insight, without a compelling connection to objectives, and without insightful comparison with other variables. There appears to be some confusion over the use of the terms *analysis* and *assessment* in COIN and IW, at least in some of the literature of the last decade. Absent clear definitions, and at least a clear understanding of the difference within the analytic community, it has been hard for many commanders, staff officers, and analysts to define the roles of analysis and assessment, and to ensure they are used most efficiently and effectively.

One passage stood out during the literature review as being particularly relevant, succinct, and practical, so we have replicated it here as an exemplar of existing lessons learned for analysis in OIF and OEF. The 2010 MORS working group led by COL Thomas Cioppa (USA) produced a series of recommendations for improving analytic support to the warfighter in IW/COIN. This working group report included input from modelers and simulations experts. While the report focused specifically at the tactical level and on assessment, it is relevant to all levels of warfighting. It is worth noting that the recommendations in this passage are intended specifically to inform ORSAs and commanders employing ORSAs, but most appear to have general relevance for IW analysis, and perhaps indirectly for modeling and simulation. They addressed several pertinent issues, as shown in the following quotes:[84]

Analysts should be vested in the problem.

> There are many best practices that can be employed at the tactical level. A key overarching concept is that the analyst should be vested in the problem, meaning they need to understand the environment, understand the commander, and be able to give actionable recommendations.

Know how and where to find data, and compare data.

> [Analysts] must be versed in the ability to know where data resides and quickly get it from one network to another so information can be shared. The analyst cannot rely on one piece of evidence, but must look for multiple sources. This again implies the analyst should speak to others within the staff (not only internally, but higher and lower) to see how they are viewing a problem.

Analysts must answer the question asked.

> The analyst must answer the question. Do not answer the question you want, but answer what is asked. If you do not think it is the right question, then clarify prior to doing the analysis.

---

[84] Cioppa, Eggen, and Works, 2010, p. 7.

Help commanders understand complexity.

It is acceptable to have a simple approach to a problem. The problems are complex, but a solution may not have to be.

Build credibility as an analyst.

It is up to the analyst to build credibility. If a previous analyst has built up a reputation, it is your job to make it better.

Know the players, use all available tools—both art and science.

The analyst must understand capabilities and vulnerabilities of friendly and threat forces. Ensure that you employ a structured problem evaluation and use both art and science of OR so that you can operationalize results that speak to the warfighter. The analyst must have a good understanding of ArcGIS and Microsoft Office, but should also have working knowledge of at least one statistical software package. Using the ORSA toolbox is of significant benefit.

Partner analysts to improve effectiveness and efficiency.

There should be consideration to partnering ORSAs and not having them work as individuals. Each analyst has strengths and weaknesses and teaming them gives excellent results.

CHAPTER THREE

Support to Force Protection Decisions

This is the first of four chapters addressing each of the four categories of decisions identified by our inductive analysis of the literature and interview results. This chapter includes a brief description of the first decision category—force protection—followed by a reflection of commanders' insights, derived from our interviews with commanders at the tactical, operational, and strategic level of command. Next, we include several examples of how analysis and M&S supported the effort to protect the force from IEDs. All of these examples were taken from the analytic support to the JIEDDO from late 2005 through 2008. The Core Operational Analysis Group (COAG)—consisting of several of the Federally Funded Research and Development Center organizations and other government and nongovernment organizations—was created specifically to address the IED threat. Although force protection is more than C-IED operations, most commanders agreed that this was clearly the most stressing problem facing their commands and therefore demanded the most analytic support. All of the examples were drawn from OIF and are meant to exemplify challenges, methods, and issues associated with countering the IED threat before it is detonated; i.e., "left of the boom."

In the context of COIN and IW, force protection includes actions taken to prevent or mitigate hostile acts against U.S., coalition, and host-nation forces, resources, facilities, and critical information. These actions are intended to conserve the force's ability to protect the population and conduct offensive operations in support of host-nation objectives. Force protection can also entail offensive operations against known or suspected enemy forces and encompass a range of analytic support for broader decisionmaking. This chapter focuses primarily on analytic support to what might be considered defensive decisionmaking. All-source intelligence analysis appears to be the primary method of support for offensive operations.

Commanders' Decisions in Force Protection

Decisions affecting force protection tend to address direct fire, indirect fire, or IED threats. Strategies for dealing with these threats can generally be categorized as either preparation for the consequences of an event, or activities to prevent the event from occurring. These are typically divided into what are known as "right of boom" and "left of boom" actions, referring to whether the action takes effect prior to an attack as opposed to after or during an attack.[1] Analysis might support protective actions such as the hardening facilities (e.g., HESCO barriers),[2] personnel (e.g., the Improved Outer Tactical Vest), or vehicles like the Mine-Resistant Ambush Protected vehicle. Preventive actions include the interdiction of enemy forces, an improvement in situational awareness, or the avoidance of a threat. Prevention might also include aerial interdiction of an IED emplacer, the emplacement of security towers to avoid surprise by a massing enemy force, or the use of irregular patrol patterns (or increased or reduced patrolling) to complicate the enemy's planning efforts. Figure 3.1 depicts the commander's force protection decisions.

Figure 3.1
Commanders' Decisions in Force Protection

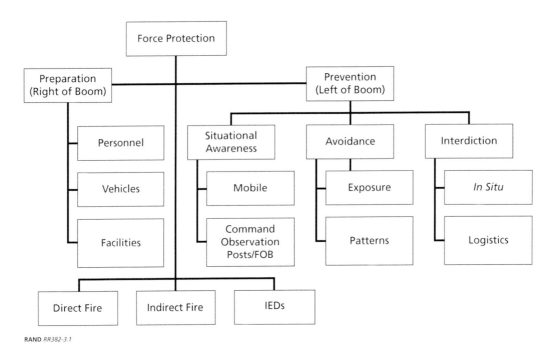

---

[1] Many activities discussed here could be considered force employment actions. However, they are taken with the explicit objective of protecting the force. Force employment decisions excluded from this discussion include those that are not motivated by force protection considerations (e.g., attacking the network is included, but disruption operations are excluded), and decisions that fall within the traditional use of maneuver forces within a direct-fire engagement (e.g., close air support, quick reaction force).

[2] The HESCO bastion is used for flood-control and military fortification and the name of the British company that developed it in the late 1980s. It is made of a collapsible wire mesh container and heavy-duty fabric liner, and used as a temporary to semipermanent barrier against blasts or small-arms fire. It has seen considerable use in Iraq and Afghanistan.

Direct and Indirect Fire

Commanders placed relatively little emphasis on analysis to counter either direct or indirect fire threats, perhaps feeling that dealing with those threats fell within their traditional core competencies. That said, a commander's emphasis often changed depending on the phase of the campaign and enemy activities.

One commander noted that upon entry into a new AoR, insurgents would initially engage coalition forces with direct fire (DF) assaults, but over time the DF threat declined and the IED threat became the principal concern.[3] Many of the commanders interviewed speculated that force protection for FOBs and COPs might be a promising application for M&S capabilities (e.g., determining what COP position would minimize dead space and blind spots).[4] Commanders did value tracking of direct and indirect fire (IDF) trends in the context of campaign assessments, discussed in Chapter Six.[5] The following vignette is a good example of some of the work done to reduce the IDF threat using "left of boom" analysis.

IEDs

Unsurprisingly, the force protection issue of greatest concern to commanders in OEF and OIF was the IED threat. Commanders wanted analysts to keep them apprised of IED trends and patterns (including where, when, what kind, and how often), identify what factors drove IED trends and where IEDs were originating from (caches, logistics, and financial networks), and predict when and where IEDs would be emplaced in the future. In one case, a commander at the Brigade Combat Team (BCT) level focused his ORSA entirely on C-IED analysis.[6]

Commanders used "left of boom" C-IED analysis to inform decisions about force protection profiles, electronic warfare assets (e.g., Warlock units), patrol and logistic convoy routes and schedules, "hunter/killer" team and checkpoint locations, as well as route clearance package and ISR allocations.[7] Commanders appreciated analysis that predicted the location of IEDs, but felt there was still a need for additional capability to predict where and when IED emplacers would be active.[8] Some analysts reported that commanders were so focused on C-IED that they were reluctant to place addi-

---

[3] Interview with commander.

[4] Interview with nine commanders.

[5] Interview with nine commanders.

[6] Interview with deputy BCT commander. This deputy commander said that if allocated another ORSA, the next priority would have been analysis of the population, which they had minimal capacity or capability for on this deployment.

[7] Interview with commander.

[8] Interview with commander. The level of confidence commanders had in C-IED analysis was surprising, given that they continue to be a major cause of U.S. casualties. One commander noted, "We knew where the IEDs were so we could avoid them, or engage them if wanted to get into trouble."

tional reporting burdens on subordinates, even in instances where it might have played significant dividends in supporting C-IED analysis. In the case of one Theater Special Operations Command, this challenge (for a less urgent reporting requirement) was elided by first placing reporting requirements on support elements rather than maneuver elements, allowing an opportunity to prove the value of the proposed analysis to the commander.[9]

The Basic Analytic Problem

The mandate to develop methods to defeat the IED "left of the boom" required a fundamental understanding of how an IED is fabricated, moved, detonated, and financed. Early discussions of the problem (the decisions that needed to be made) resulted in the development of the IED event chain depicted in Figure 3.2.

An interesting feature of this event chain is that understanding the activities in each of the events in the chain requires intense intelligence activity. Consequently, the analysis conducted by the COAG supported intelligence and not just operations.[10] Nevertheless, the intelligence decisions or questions were clear: How are IED operations funded? How are emplacers and assemblers of IEDs recruited? Where and what kind of materials are gathered and used in making IEDs? Where are the devices assembled? How are they moved from assembly location to site of emplacement? Where are they stored (cache locations) prior to use? How are they detonated and what are the likely targets? Post detonation analysis consisted mainly of forensic and biometric analysis aimed at identifying the type of device used and the possible emplacer (crime scene analysis).

Figure 3.2
The IED Event Chain

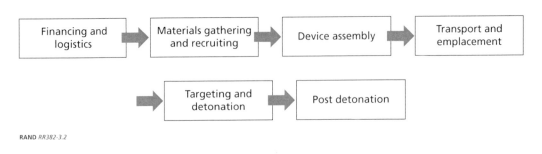

RAND RR382-3.2

---

[9]   Interview with analyst.

[10]   This idea is elaborated on in Walter L. Perry and John Gordon IV, *Analytic Support to Intelligence in Counterinsurgencies*, Santa Monica, Calif.: RAND Corporation, MG-682-OSD, 2008.

Analytic Support to C-IED Operations in Iraq

The IED was considered the greatest threat to the coalition forces in Iraq, and considerable time and effort were dedicated to countering it. First, right-of-boom measures were taken such as up-armoring vehicles and developing radio frequency jammers to disable remotely detonated IEDs. With the advent of the Explosively Formed Penetrator, more sophisticated methods were required. The penetrator is triggered by an infrared sensor detecting heat from a vehicle. Sophisticated microwave jammers were then required to disable the detonators and heat attractors were used to deflect the effects of the penetrators.

Increasingly sophisticated protective measures were taken as time went on. The organization charged with defeating the IED was JIEDDO. It was created in fall 2006 as the successor to the Joint IED Defeat Task Force, which was charged with the same mission. As JIEDDO was beginning to organize, it became clear that initiatives taken to defeat the IED must include left-of-boom activities as well, and this called for an analytic effort that, at the time, was considered to be on the order of the effort to defeat the German U-boat campaign in the North Atlantic in WWII. To tackle the problem, JIEDDO created COAG, consisting of a consortium of several organizations: the Institute for Defense Analyses, the Center for Naval Analyses (CNA), the RAND Corporation, the Applied Physics Laboratory (APL) at Johns Hopkins University, MITRE Corporation, George Mason University, TRAC, the U.S. Military Academy at West Point, and the Joint Forces Command. This group was charged with conducting analyses with existing data, aimed at defeating the IED left of the boom.

A Brief Introduction to Attack the Network

Because analysts working on force protection problems tended to focus on left-of-boom preventative measures, their analyses often revolved around defeating insurgent networks. This kind of network analysis is one of the crossover points between OA and intelligence analysis. Deployed ORSAs involved in force protection in both Afghanistan and Iraq provided direct support to intelligence and operations staffs; in this chapter we present both deployed and reachback vignettes describing these kinds of analyses. Analysis of insurgent networks could take many forms, but over time both analysts and commanders began to refer to these efforts as Attack the Network (AtN) analyses.[11] In simple terms, AtN "consists of identifying [a network], determining whether it is important or not, and using the means at hand to defeat it."[12] This is in many ways similar to the 1990s and early 2000s-era concept of System-of-Systems Analysis that remains prevalent in current doctrine. Some argue that the AtN theory can be expanded to encompass all actions—including nonkinetic actions—taken by

---

[11] This term began to appear in various official documents in the mid-2000s, and was eventually incorporated into various technical pamphlets and doctrinal publications.

[12] Kinner, 2012, pp. 4–7.

a military force in any operational context.[13] Regardless of whether this is the case, tactical-level AtN analysis was one of the most common force-protection approaches used by ORSAs in both Afghanistan and Iraq.

Figure 3.3 shows how JIEDDO envisions the functioning of a generic insurgent IED network. The insurgents conduct targeting, surveillance, planning, and then conduct an attack with an IED. They observe many of these attacks, learn from the results, and sometimes incorporate these lessons into the next cycle of attacks. The purpose of AtN in this context is to identify the people involved in the cycle, determine the timing and geographic locations of various actions within the cycle (e.g. movement of a device), and recommend actions to disrupt the cycle or eliminate the network.

It would be rare for either an ORSA or an intelligence analyst to be able to see and understand all parts of an insurgent network, and in many cases knowledge was based on information of questionable reliability. However, even with incomplete or partly inaccurate knowledge, it was often possible to target sections of an insurgent IED or

Figure 3.3
JIEDDO Conception of an Insurgent IED Network

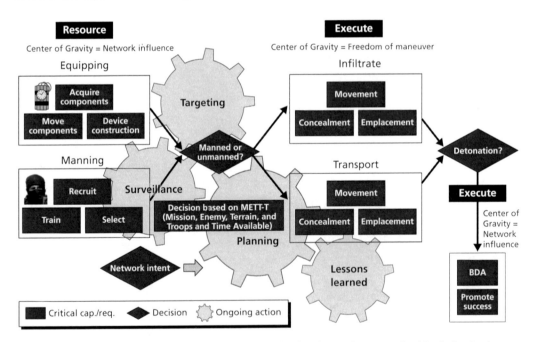

SOURCE: Frederick Gaghan, "Attacking the IED Network," briefing, Joint Improvised Explosive Device Organization, May 5, 2011, p. 10.
RAND RR382-3.3

---

[13] Kinner, 2012, pp. 4–7.

IDF network, thereby disrupting the cycle. The following brief vignette describes one of these successful AtN analyses.[14]

Vignette: Tracking an IDF Shooter Using Overhead Sensors

Between 2001 and 2012, analysts in both Afghanistan and Iraq learned to take advantage of the improving capabilities of unmanned aerial systems (UAS) to improve their analyses. In cooperation with systems managers and intelligence collections specialists, analysts worked to "cross-cue" various systems in order to track insurgents for force protection analyses. In this case an analyst deployed to Iraq leveraged a combination of assets—RADAR and UAS—not only to identify a threat but to eliminate that threat with a kinetic strike. The analyst—an ORSA—was responsible for force-protection analysis along the main supply route (MSR) that ran through a unit's area of operations. The analyst was able to identify both IEDs and a number of IDF points of origin, from which a particular shooter was launching attacks against coalition bases. A combination of patience and careful analysis allowed the analyst to play a key role in bringing these attacks to an end:

> We tracked a single vehicle for a week, on and off, and eventually were able to kill an IDF shooter in the act. It turned out that this one shooter was the key to the whole IDF system in the region. Attacks dropped off to almost zero for six months.[15]

This analyst viewed force-protection analysis as a systems-oriented task that provided insight for both intelligence and operations sections. The goal was to find a "minimal cutting plane at which we separated the shooters from the leaders, or cut off the resources from the shooters." This is quintessential AtN analysis, and this analyst and others we interviewed felt this was a very successful approach for tactical force protection.

Vignette: COAG Analytic Efforts (2005–2008)

We chose this extended vignette to exemplify the broad range of analytic work done to support force-protection efforts in the field from 2003 through 2012. This particular project involved both reachback and "down range" support, it resulted in the development of both models and simulations that were used to inform commanders' decisions, and it involved the work of multiple partnered organizations and agencies. In full disclosure, RAND was one of the primary organizations involved in this effort. Therefore, we are highlighting some of our own work. However, we are also highlighting the work

---

[14] Interview with analyst.

[15] Interview with analyst.

of analysts at the Center for Naval Analyses, the Institute for Defense Analyses, and elsewhere, and we are able to provide useful insight into the methodologies developed and employed.

This vignette also includes a description of the work conducted in 2005 in support of the Joint IED Defeat Task Force, before the establishment of the COAG. Many of these studies were published and citations are included where appropriate. What follows are analytic efforts that address various activities in the IED event chain. This is not a complete list, but rather one that illustrates how operational analysis can be used to support intelligence operations.

MOEs for the Counter-IED Fight

Although not explicitly part of the IED event chain, the effective assessment of how well C-IED operations are performing is a critical component of both intelligence and operations. This first research effort was designed to develop an assessment methodology that could be applied to the C-IED fight.

This work was in response to the need for a set of MOEs and supporting analysis for the C-IED campaign in OIF. Senior military and civilian leadership expressed concern about how well the IED fight was going. This spawned a series of questions such as: Are the C-IED investments in technology; systems; and associated tactics, techniques, and procedures (TTPs) paying off? Are they cost-effective? Is the action/reaction C-IED battle converging in our favor, or does the adversary continue to adapt effectively? This work consisted of describing the steps that must be taken to build a foundation for the establishment of MOEs to include the desired end state, strategic objectives, the measures, and finally the associated metrics. Although the recommendations from this work were not implemented directly, the MOEs and metrics were discussed in terms of trends briefings but without the benefit of diagnostics—an analysis of the observed trends.

Sustaining Intelligence Directed Action

Detecting emplacers in the act in order to take some action became a central focus of C-IED operations. If ISR assets detected an emplacer, then the coalition had a decision with three options: Kill the emplacer, monitor his activities after emplacement to gain intelligence on where his cache was hidden, or alert a quick reaction force to seize the emplacer. Analysts examined the application of ISR in the fight and defined the functions and capabilities that are required to ensure the right information is collected, analyzed, and effectively applied. An implementation model for building the structures and systems needed to gain as much advantage as possible from ISR activities was proposed. MOEs in ISR efforts—which have operational importance for the day-to-day evaluation of sources of information as well as a broader value in assessing ISR efforts as an element of the overall military effort—were also suggested. The next study implemented the conceptual model to address detection of emplacers along the airport road in Baghdad.[16]

Reconnaissance, Surveillance, and Target Acquisition Options for the IED Fight

At the time this research was conducted, the airport road from the Green Zone (secure area) in Baghdad to the airport came under almost daily attack by IEDs. The command was interested in determining what could be done to oversee the road using only assets available in that country at the time.

The study provided quantitative insights on the relative merits of several reconnaissance, surveillance, and target acquisition (RSTA) options available to the command. This was done primarily through the use of computational analyses and a force-on-force simulation using the Joint Conflict and Tactical Simulation (JCATS).[17] The computational analyses characterized the sensitivity and discrimination capabilities of several sensor packages while the simulation was intended to represent appropriate sensor/platform combinations operating in a stressing route reconnaissance vignette. The area chosen for the scenario vignette was the seven-mile road between the Baghdad airport and the Green Zone. Analysts used the digital terrain data for the airport road to portray a representative set of surveillance situations that simultaneously had all the elements needed to compare different technology options.

Digital terrain and building data from the area was put into JCATS, and systems such as aerostats, tower cameras, UAVs, and sniper teams were inserted into a representative scenario. Complicating features were also present, such as milling noncombatants and civilian vehicle traffic. The constructive simulation was able to compare the different RSTA options on the basis of coverage, resolution, detections, identifications, time delays, and vulnerability to enemy fire. As shown by the sample screen shot in Figure 3.4, the system was unable to realistically represent clutter, posture, crowd characteristics, or complex behaviors. All of these can be critical to spotting and reacting to enemy threats. It was found that a combination of systems was needed to maintain effective surveillance of the road, as UAVs provided high-resolution but sporadic coverage, tower cameras and aerostats were often blocked, and sniper teams were vulnerable.[18]

---

[16] Details of this study are included in Walter L. Perry, "Linking Systems Performance and Operational Effectiveness," in Andrew G. Loerch and Larry B. Rainey, eds., *Methods for Conducting Military Operational Analysis*, Alexandria, Va.: Military Operations Research Society, 2007.

[17] JCATS is an entity-level constructive training simulation system that provides command-level training. It has a digital interface and the capability to simulate the joint battlespace and Military Assistance to Civilian Authority missions.

[18] This is derived from research conducted by Randall Steeb, Jon Grossman, Morgan Kisselburg, and Manuel Carrillo for RAND in 2006.

Figure 3.4
Sample JCATS Simulation Screen Shot (Aerostats Sensing Entities Along the Airport Road)

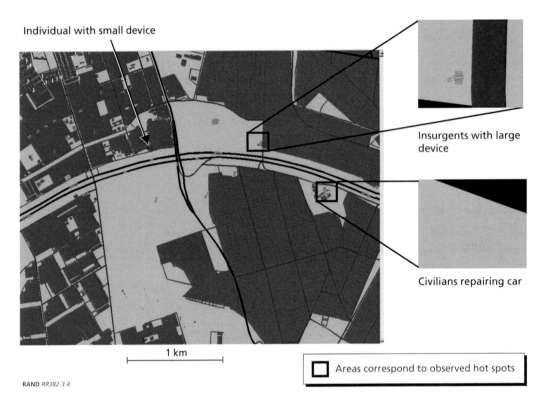

RAND RR382-3.4

The Actionable Hot Spot Resource Planning Tool

One of the critical intelligence information requirements for commanders in Iraq was the location of emplaced IEDs. In most cases, commanders simply chose to avoid known or potential emplacement areas, but on occasion, when sufficient corroborating intelligence was available, they chose to watch the area in hope of capturing an emplacer or monitoring his activity. The Actionable Hot Spot tool (AHS) was developed to identify these areas on a daily basis.

AHS uses recent history of the time and place of IED-related activities to detect clustering patterns that *may* be indicative of future threat activities in the *immediate* area. The AHS was tested on past data in support of six brigades in Iraq from August to December 2006. It produced variable but often encouraging results. JIEDDO's Counter-IED Operation Integration Center evaluated the tool in anticipation of offering it to units in Iraq. That evaluation was completed in March 2006. A subsequent evaluation found that:

According to the reviewers, the proposed AHS methodology combines several well-developed data mining approaches and may work well under certain conditions. However, at the present time, the technology has several serious limitations with regard to developing and deploying real world production applications. These limitations include scalability, parameter selection problems and lack of readiness for production computing environments.[19]

Nevertheless, JIEDDO decided to install AHS at the center, provided that RAND supply an analyst on site to install the tool and conduct a more rigorous test using current data. Where clustering occurs in an area of operations, results are good: If clustering does not occur or if IED attacks fall off, then results are poor.

The field test in 2006 consisted of support to six brigades of the 4th Infantry Division in Baghdad at the time and one logistics unit. Over the period of the test, the tool proved to be accurate 30 percent of the time—an improvement over the 5 percent accuracy obtained without the tool.[20]

Counterbomber Targeting

In fighting irregular forces, it is often the case that the friendly forces know very little of the composition of the enemy force, its command and control structure, its objectives and TTP. The IED event chain is a case in point. To gain information concerning materials and technologies used in construction of IEDs—and of the network that recruits emplacers, assembles the devices, and delivers them to the target—it is critical to understand the nature of the enemy network, including how it spreads the technology required to construct IEDs.

Two promising methodologies were developed: *knowledge migration* analysis and *social network* analysis. The former combines engineering and geospatial analysis to map the migration of IED triggering technologies across Iraq, and it associates IED cells with the spread of these technologies. The latter was completed in association with the National Security Agency's Social Network Analysis Work Center. It consists of analyzing reports produced by the intelligence community in Iraq to extract and visualize social relationships between insurgents and insurgent groups. Knowledge migration was used to trace the migration of two trigger types: long-range cordless telephones and personal mobile radios. Social network analysis was used to produce a comprehensive depiction of both Sunni and Shi'a IED networks. The National Security Agency merged the field-generated intelligence with their intelligence sources to update and expand this network.

---

[19] *Proposal Review: Predictive Analysis Tools Assessment, Actionable Hot Spot Monitoring (AHSM), Building Time-Sensitive Clusters in Time and Space*, PATA Technical Review, undated.

[20] This is derived from research conducted by Ryan Keefe, Katharine Hall, and Adrian Overton for RAND in 2007, and also from Ryan Keefe and Thomas Sullivan, *Resource-Constrained Spatial Hot Spot Identification*, Santa Monica, Calif.: RAND Corporation, TR-768-RC, 2011.

Computation and Simulation Test Bed

Suggestions for gaining intelligence of the IED networks in Iraq were encouraged by JIEDDO early in the C-IED fight. Suggestions from private citizens, industry, and government organizations were plentiful. The problem became determining which were worth pursuing. This led to the idea of a general test bed for promising ideas.

At the time, JIEDDO received upward of 30 proposals for new C-IED equipment each week. Most new ideas could be evaluated rather quickly or dismissed as duplicative. However, for the more costly equipment proposals that required advanced technology, JIEDDO required an evaluation. A computation and simulation test bed was developed to evaluate alternative C-IED proposals. The simulation team first examined the physics and engineering, then simulated the proposed equipment in both an urban and rural synthetic environment using the JCATS simulation platform. Their major efforts involved (1) completing an assessment of competing counter-PIR (passive infrared) triggering devices, and (2) an evaluation of alternative command-wire triggering devices. Their counter-PIR recommendation was instrumental in selecting the winning technology.[21]

Gaming Blue-Red Interactions

An important piece of intelligence in the C-IED fight is understanding enemy motivations and their operational and strategic objectives. Although not on the IED event chain, understanding enemy goals and objectives (as we have seen with friendly goals and objectives) is the first step in assessing how well the insurgent group is achieving them. One way to study enemy goals and objectives is to apply game theory to the Blue-Red interaction assessment problem.

The use of game theory to analyze military operations is not new, and it was only natural that we examine its applicability to C-IED operations, or to be more specific in this case, friendly-enemy interaction analysis. Insurgent elements (Red) make their own decisions about emplacing IEDs, choosing when and where to emplace, the triggering device, and when to detonate the IED; the set of Red strategies is in correspondence with the set of possible answers to these questions. In general, the success of a Blue mission, and the outcome of a Red IED attack, depends on how well "matched" Red's strategy is to Blue's strategy. Red must attack when and where Blue will travel, and may need to adjust its tactics in a way that is tuned to the given Blue mission. The analytic team assumed that the outcome of the game—or the fate of the mission—could be measured in terms of an expected payoff thought to be derived from the consequences of Red propaganda (sometimes referred to as the "CNN effect"), friendly casualties, etc. Crucially, the analysis did not depend on actually measuring the payoffs. One approach was to examine relative payoffs. For example, Red would

---

[21] This is derived from research conducted by Randall Steeb, Jon Grossman, Morgan Kisselburg, Jeffrey Sullivan, Ravi Rajan, and G. Heath for RAND in 2006.

conclude that it had better achieved its objective with more Blue casualties than fewer. The assumption is merely that the payoffs *could be* evaluated on some ordinal scale.

In the example depicted in Figure 3.5 (drawn from the study), both Red and Blue must choose a route: Blue to travel from point A to point B, and Red to decide where to emplace an IED. Figure 3.5 illustrates the "game board." Red's choices are the columns and Blue's are the rows. We first note that the game is not zero-sum; i.e., it is not the case that Red's loss is always equal to Blue's gain and vice versa. In fact, we have that Red is indifferent to, or gives different value to, Blue rewards (Red costs) and Blue is indifferent to, or gives different value to, Red rewards (Blue costs). The rewards and costs are illustrated in each cell. If Blue chooses Route A and Red does the same, then Red expends resources to emplace the IED and gets a propaganda advantage from the detonation of the IED and Blue suffers casualties and fails in its mission to get to point B.

Development of a Platoon-Level Simulation Environment

In response to concerns that C-IED analyses focused more on the operational and strategic level, a simulation and computation environment that could be used to explore coalition force activities and threat IED responses at the platoon level was developed. The platoon level modeling task consisted of four research activities: (1) assembling information on both the systems that patrols are equipped with and the threats they are likely to counter; (2) modification of existing scenarios and the possible development of others to represent and stress patrol systems and TTP; (3) computing the effects of Red counters on the patrol's performance using science and engineering prin-

Figure 3.5
Choosing a Route

	Route A	Route B
Route A	**RED Costs and Rewards** • Resources to emplace • "CNN Effect" **BLUE Costs * Rewards** • Coalition casualties • Mission failure	**RED Costs and Rewards** • Resources to emplace **BLUE Costs * Rewards** • Mission accomplishment
Route B	**RED Costs and Rewards** • Resources to emplace **BLUE Costs * Rewards** • Traveling less-preferred route • Mission accomplishment	**RED Costs and Rewards** • Resources to emplace • "CNN Effect" **BLUE Costs * Rewards** • Coalition casualties • Mission failure • Traveling less-preferred route

RAND *RR382-3.5*

ciples where applicable; and (4) simulating the tactical operation of the patrol in an operational environment as it performs its mission while reacting to Red counters. In March 2008, several simulations were linked with 3-D visualization tools and interactive systems.

A typical scene from the demonstration is shown in Figure 3.6. The screen shot shows a typical built-up urban area, with roads, foliage, trash, and local vehicles. The graphics engine produced extensive texturing, shading, and detail, and controllers could observe the scene from many different vantage points, such as the driver, passenger, and gunner's positions, along with the "stealth" or third-person mode. In the demonstration, three man-in-the-loop stations were present that could be configured to control either friendly or enemy systems. Additional vehicles could be controlled with computer-generated forces, or they could be assigned to JCATS.[22]

Figure 3.6
Urban Scene from Virtual Simulation

SOURCE: Forterra Systems, OLIVE (On-Line Interactive Virtual Environment).
NOTE: In commercial gaming, the Simulation and Technology Training Center and Forterra
have demonstrated 80–200 moving avatars, each controlled by man-in-the-loop stations
connected over the Internet.
RAND RR382-3.6

---

[22] This is derived from research conducted by Randall Steeb, Jeffrey Sullivan, Morgan Kisselburg, Jon Grossman, Daniel Williams, and Catherine Kuehne.

Summing Up

Commanders in OEF and OIF needed time and space to set the conditions for stability, a condition that generally took years to achieve even at the local level. Commanders at all levels needed to ensure not only that they suffered minimal losses to their own troops in order to keep sufficient manpower on hand, but also that there was adequate freedom of movement for their forces, for host-nation forces, and for civilians. Simple improvements in armor or anti-IED technology such as jamming devices were insufficient for their needs. Force protection therefore took on an expanded meaning in the contexts of OEF and OIF, becoming at times an all-consuming analytic effort that involved models, simulations, and extensive and detailed operations analyses.

The IED event chain served as the backdrop for the wide array of analyses across DoD and in the civilian research community supporting operations in OEF and OIF. Most of these were operations analyses in support of broader intelligence analyses; these in turn informed commanders' decisions regarding the allocation of resources for the C-IED fight and for force protection options. The tremendous focus placed on C-IED led to the development of a range of new analytic methods, models, and simulations, some of which are described in this chapter. These analyses led to some clear successes at the tactical level, and in Iraq they did help give commanders sufficient time and space to set conditions for at least a tenuous stability. Modeling, simulation, and analyses in support of tactical AtN efforts were innovative, and both deployed and reachback efforts could often demonstrate clear value.

CHAPTER FOUR

Support to Logistics Decisions

Several of our interviews and the documents we reviewed consisted of discussions centering on how models, simulation, and analytic methods were used to support logistics decisions. The central themes seemed to be controlling the movement of supplies to include convoy control, trade-offs in lift capability, developing alternative withdrawal schedules from Iraq, location of specialty surgical teams, and identifying medical evacuation requirements in Iraq.

In general, the use of M&S and analysis to support logistics decisions has been successful. The cases we discuss in this chapter (and others we do not include) are more amenable to traditional operational analysis techniques than some of the other IW topics, such as campaign assessment and force protection. Data supporting decisions concerning convoy routes lift capability, surgical team location, and medical evacuation requirements consist of tangible items such as road networks, vehicle lift capacity, troop sizes, and air ambulances required to support deployed troops. In these examples, models were used to answer specific questions—some in-country and others utilizing reachback centers.

Commanders' Key Decisions and Challenges in Joint Logistics

This section presents commanders' views on logistics challenges in OIF and OEF that have benefited or might benefit from modeling, simulation, and analysis. Though some of the challenges encountered in OEF and OIF were unique to those theaters (e.g., effect of altitude on vertical lift support in Afghanistan), some are generalizable across the broader spectrum of IW cases. U.S. forces conducting COIN in OIF and particularly in OEF have been operating in a far more distributed fashion than for which they were designed.[1] What might be termed "extended distributed operations" create long supply chains, thinly spread logistical assets, and in many cases, logistical deficits. In Afghanistan, environmental issues compound the logistics challenge. Dispersion,

---

[1] Headquarters, U.S. Army, "Army Modified Table of Organization and Equipment (MTOE) Scrub Strategic Issues," Memorandum, Department of the Army Military Operations FMF, December 15, 2011b.

threat conditions, poor and nonexistent road networks, and sharp terrain contouring place a high demand on vertical lift. In turn, this high demand increases maintenance requirements and burdens. One commander noted that his battalion-sized unit might be allocated one airdrop to support 56 small outposts, placing a significant redistribution requirement on his limited organic assets.[2]

Many commanders and staff highlighted the importance and challenge of managing logistics in irregular environments and felt more analytic support could be valuable, though that view wasn't universally held.[3] Managing logistical requirements was seen as consuming a substantial amount of staff time, and it demanded that individual staff members and sections demonstrate innovative thinking.[4] Many commanders noted that logistics is a crosscutting issue that affects many of the other areas discussed in both the force-structuring (e.g., request for forces) and force-protection section of this report. Though none of the commanders or staff we interviewed reported having to abort operations due to logistical constraints, there was considerable concern over the effect of "feast or famine" supply chains on their tactical units: They had either a significant excess or a significant deficit of supply, creating a disruptive ripple effect across both planning and operations.[5] As a result, commanders and staff were interested in additional analytic support to address two core logistical challenges: forecasting demand and optimizing use of resources. Addressing these challenges is complicated by three additional factors: force protection, terrain and weather, and asset visibility. Figure 4.1 depicts commanders' key decisions and concerns for logistics.

Forecasting Demand

Irregular warfare is dynamic: Mission requirements shift over time or unexpectedly at any one time. Because IW/COIN operations tend to last for several years, logistics planners try to maximize efficiency over the course of the campaign. This means establishing a rhythm of supply, maintenance, and other services that simultaneously meet demands while minimizing costs, wear and tear on assets, and impact on personnel. It would be possible to optimize this rhythm or find an equilibrium if IW/COIN operations were static, but logisticians must function in the same operational environment as their commanders. The "feast or famine" dynamic is driven by changes in supply requirements as operations evolve—realities of IW keep the system out of equilibrium.[6] Logistics staff hoped for some capability that would help them forecast logistic require-

---

[2] Interview with commanders.

[3] Interview with commander. One Special Operations Forces commander felt the logistics enterprise had sufficient flexibility to adapt to whatever scheme of maneuver he adopted with its current staff and analytic capabilities.

[4] Interviews with analysts.

[5] Interview with analyst.

[6] Interview with analyst.

Figure 4.1
Commanders' Decisions in Logistics

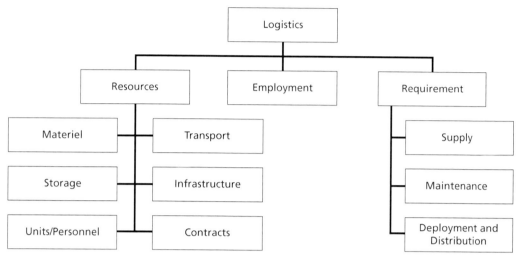

RAND *RR382-4.1*

ments. In addition to their own logistical requirements, some commanders needed to plan how to ensure their partnered host-nation forces would remain sufficiently supplied to conduct operations.[7] Requirements growth might be readily projected based on known operational plans, but the dynamic impact of the insurgents, the population, host-nation officials, and U.S. policy shifts cannot be accurately predicted.

Planning and forecasting efforts were complicated by a lack of asset visibility in the supply chain. Different commodities and classes of supply were tracked in different systems; some staffs were unable to identify projected arrival times. At times, DLA would shift suppliers or quality would decline, so planning factors for a given commodity would change.[8]

Managing Resources

If forecasted demand outstrips available resources, commanders must decide to request additional resources, adjust how those resources are employed (an optimization problem), or adjust the concept of operations for their force (i.e., reduce demand). Commanders viewed the last option as the least desirable. Much of the time, the feast-or-famine supply dynamic was driven by the availability of specific classes of supply (e.g., fuel management), but transport assets were reported to be the primary chokepoint.[9]

---

[7]   One interviewed staffer described ANSF logistics as a "disaster."

[8]   Interview with analyst.

[9]   Interview with commander.

Indirectly, limited transportation infrastructure drove the need for particular transportation assets (e.g., lack of hard roads or airfields). Terrain and weather could unpredictably restrict the times when a unit could be resupplied, disrupting resupply plans. Seasonal effects sometimes had a dramatic impact on the feasibility of ground distribution networks. Units at times had to make use of airdrops from C-130s when their planned resupply routes became inaccessible.[10]

Staff hoped that additional analytic support might help them optimize logistical plans (e.g., routing, transport platform mix), or even identify best practices from across the theater. This interest extended beyond the direct management of organic coalition assets to the performance of contractors (local national and international) providing logistical support.[11]

Vignette: Theater Airlift

The basic problem centered on looking for ways to free up rotary-wing aircraft being used for troop transport by substituting fixed-wing transport.[12] During the surge in Iraq in early 2007, Army CH-47 helicopters were being used to transport personnel. With the increase in personnel from 138,000 to a peak of 158,000 personnel, the number of "blade-hours" for all CH-47s increased to 70 per month. The decision to be supported then was finding alternate transport that would bring CH-47 blade-hours down to a manageable 58.

In addition to this excessive use, CH-47s were also needed to support the war in Afghanistan. At the time, the CH-47 was the only helicopter in the U.S. military inventory that could clear the mountainous terrain and touch down at bases to deliver personnel and equipment to isolated locations. This presented an additional reason to look for some substitute for the CH-47 in Iraq. Thus, the decision to be supported expanded to finding alternate transport that would reduce CH-47 blade-hours to 58 and free up several machines for deployment to Afghanistan.

Analysis
The basic problem centered on determining the effectiveness of C-130 aircraft when augmenting CH-47 missions in Iraq. First, the research team studied the CH-47 operations in Iraq (the requirement). This was followed by determining which of these missions could be performed by C-130s. This second task determined the number

---

[10] Interview with commander.

[11] Interview with analyst.

[12] This section is based on interviews at the TRADOC Analysis Center (TRAC) at Ft. Leavenworth Kansas and D. Anderson, D. Henderson, A. Hummel, R. Spivey, and J. Wray, *Intra-Theater Air Lift Planning—Redux*, TRAC-L-TR-10-044, TRADOC Analysis Center, July 2010.

of blade-hours that might be saved by the C-130 augmentation. The approach the team took was to examine the effect of substituting a C-130 for each two-ship CH-47 formation.

By way of illustrating the analytic method, the team selected nine destinations for air passengers and depicted the CH-47 "ring-route" used to deliver passengers (Figure 4.2). In this diagram, the arrows depict the direction of travel. The green nodes are airfields accessible by helicopters only and the blue nodes are airfields accessible by C-130s. Assuming that each flight leg takes ten minutes for the CH-47 formation, all 42 passengers can be delivered in 90 minutes.

Next, they added a C-130 route as depicted in Figure 4.3 by the blue arrows. The C-130 moved the 28 passengers to be delivered to the second blue node. This reduced the CH-47 flying time from 90 minutes to 80 minutes and a reduction in blade-hours of 11 percent.

The model that the researchers used to analyze the problem was the Marine Assault Helicopter Planning Assistance Tool (MASHPAT). MASHPAT was developed by John Wray, a student at the Naval Postgraduate School in Monterey, California.[13] The model was designed to assist U.S. Marine Corps aviation planners in providing personnel transportation in Iraq by helping to improve Marine Corps

Figure 4.2
CH-47 Ring Route

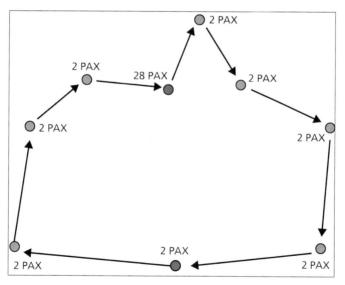

SOURCE: Anderson, Henderson, Hummel, Spivey, and Wray, 2010.
RAND RR382-4.2

_____

[13] John D. Wray, *Optimizing Helicopter Assault Support in a High Demand Environment*, Thesis, Naval Postgraduate School, Monterey, Calif., June 2009.

Figure 4.3
CH-47 Ring Route with C-130 Augmentation

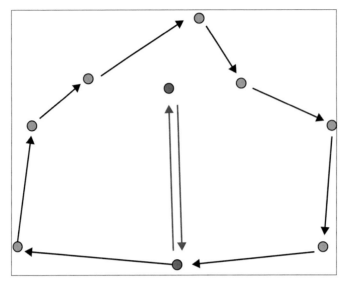

SOURCE: Anderson, Henderson, Hummel, Spivey, and Wray, 2010.
RAND *RR382-4.3*

CH-53 scheduling in western Iraq. The model produces the "best" routes to satisfy the several daily air movement requests and air support requests. It is Excel-based and was originally designed to employ the General Algebraic Modeling System to generate an optimal solution.[14] However, the problem proved to be too complex for the problem and a heuristic algorithm was developed instead.

The team executed three MASHPAT model runs for each day's number of air movement requests. The first was the baseline, with no C-130 augmentations and using the three- or two-ship CH-47 formations. The second run consisted of C-130s only, and the third featured the C-130 augmentations. This latter case consisted of replacing a two-ship CH-47 formation with a C-130, was done for all division areas in Iraq, and required that the team examine the C-130 eligible portion of the ring-route.

Data

Data consisted of inputs to the MASHPAT model. These included aircraft characteristics, airfield characteristics, and the air movement requests and air support requests (transport requirements). Central to the analysis was the requirements data. These were compiled from historical records of actual flights in all division areas of operation. From the airfield data, aircraft characteristics, and the transport requirements, the

---

[14] The General Algebraic Modeling System is designed for modeling linear, nonlinear, and mixed-integer optimization problems. A description of the system can be found at the GAMS website.

model produced estimated flight times between demand nodes and it calculated flight routes like the ones depicted in Figures 4.2 and 4.3. These were then compared to the actual times and routes as part of the model validation process.

Results

The work was completed in September 2009. The overall results depended upon the tolerance for backlogs on any given day. The team settled on an acceptable backlog of between 25 and 50; anything greater than that was construed to be infeasible. The team found that an 8 percent reduction in blade-hours was possible in the northern division area by substituting two CH-47 missions with one C-130 mission while experiencing a 50-person backlog. In the south, a 15 percent reduction was experienced with only a 25-person backlog per day by making the same substitution. Interestingly, no further improvement was possible by adding more C-130s.

These results were briefed to the Commander of the Multi-National Force Iraq. The recommendation was to substitute one C-130 for two CH-47s in the north and south but no change in the west or the Baghdad area. The cost would be a 24-hour delay for between 25 and 50 personnel in those areas respectively. The commander rejected this recommendation because by the time it reached him, it was phrased as substituting aircraft, not missions.

A subsequent study was conducted in November 2009. This time, the recommendation included the addition of two more C-130s to the theater to accommodate the increased C-130 workload. In addition, it was recommended that the study be revisited in April 2010. The reason was that base closures were rapidly increasing and a further assessment would be likely. The team did indeed update the study. With the closure of the forward operating bases, the team found that adding two more C-130s to the mix was now optimal.

Vignette: Surgery Team Placement in Afghanistan

One analyst described a study he conducted designed to answer the question: "Where should the command locate a specialty surgery team?" This work was in support of a NATO/ISAF decision. The suitability of the location selected was measured in terms of its utility in serving the deployed force. This was a fairly straightforward problem, which was subject to classical optimization methods. The biggest difficulty was determining the metrics that supported the only measure of effectiveness: the utility of the selected location in serving the deployed force. What metrics support "utility" in this case?

Analysis

As it turned out, the overriding metric had to be proximity to a field hospital. Since there were only two field hospitals in Afghanistan at the time, the possible locations for the surgery team were narrowed to the vicinity around each hospital. Clearly, this metric satisfied the "utility" measure, in that troops needing specialized surgery will likely be at one of the two hospitals.[15]

Results

Analysis selected the optimal site and the specialty surgical team was likely located at that site. (It is not known if this was indeed the case.) The point here is that this was a simple problem that was amenable to straightforward analytic techniques and operated on incontrovertible quantitative data. Consequently, a definitive result was obtained that served as the basis of a recommendation that would answer the command's question.

Vignette: Logistics Improvement to Reduce IED Exposure

A unit in Afghanistan was operating from a number of small outposts distributed across a wide swath of rough terrain. The unit was running a significant number of convoys on dangerous routes to keep these outposts supplied and the personnel fed. In turn, these convoys were subjected to risk from enemy attack and suffered significant wear and tear on their vehicles.

Analysis

"The logistics staff asked me to conduct analysis of their distribution network. I used a 'commodity flow analysis' and found that their network was limited by their use of refrigeration assets. They were running far too many convoys in order to get refrigerated goods to distant bases. I recommended installing reefers (large military refrigerator units) at these bases."[16]

Results

This solution was implemented and led to a drastic reduction in the number of convoys required to sustain these remote bases. This reduced wear and tear on vehicles, reduced hours on the road, and freed personnel up for other missions, while it also ensured sustained supply even when road conditions were not amenable to convoy travel. The analysts used some rudimentary optimization calculations to develop the solution to this

---

[15] The analyst interviewed complained that this problem should never have been assigned to the analysis cell. This is a calculation that should be a core function of the medical community.

[16] Interview with analyst.

problem, but noted, "I could have used a program like Logistics Battle Command to do this analysis. This is a program built at TRAC designed for this kind of analysis."[17]

Vignette: Using Blue Force Tracker Data to Support Convoy Security

Perhaps the most-noted concern by ORSAs was the failure of unit- and theater-level staffs to capture, maintain, and share BFT data. This is the granular data drawn from individual vehicle position reporting systems. Vehicle BFT transponders send out a great deal of information about each vehicle, its speed, its direction, its location, etc. With complete and accurate BFT data, an ORSA would be well positioned to conduct a range of optimization analyses, but also to help units move with greater safety and tactical effectiveness. For example, accurate BFT data compared to accurate IED attack data has helped analysts show commanders how to avoid IED hot spots or hunt down insurgent IED teams. At least through the late 2000s, though, these data were not being collected or shared in a way that was readily useful to analysts in the field.

Analysis

In this case, the analyst in question effectively conveyed a complete vignette in the interview with the RAND research team:

> I was told to put together an analysis of convoy operations on MSR Tampa in Iraq. The purpose of this effort was to help the theater commander determine if he could justify a reduction in security for each convoy to free up forces. I need to determine the total number of convoys that moved on the MSR from 2005–2010, and then determine what percentage of those were attacked.[18]

Once the analyst had identified the problem and scoped an approach and method, the next step was to find data to populate the simple model:

> But I didn't know any of the monthly convoy data. I actually had to get convoy movement data over the phone. I called the BFT representative at various logistics bases, but most of the data were gone. I contacted a convoy operations cell at Camp Anaconda and asked them to give me rough data on the number of convoys moving per month along MSR Tampa. These were all written data in hardcover log books, not in databases. They had to physically go through the logs to figure out how many convoys had passed through the major logistics bases from 2005–2010 and then put this into electronic form for us.

---

[17] Interview with analyst.

[18] Interview with analyst.

With data finally acquired, the analyst was able to identify at least a weak correlation that could inform the commander's decision. This quote shows how changes in the environment—in this case, the closing of a coalition base—affects the availability of data over time. The availability of data was affected by a preexisting failure to centrally capture, maintain, and distribute those data:

> I don't know how good the data were, but it showed something, it looked valid. NAVSTAR [a major logistics hub on MSR Tampa] no longer existed, and you couldn't get a hold of the data that had been collected from it.

Issues

This analyst went on to effectively describe the key issues associated with the availability of BFT data:

> Getting the problem analyzed was all about asking the right questions to the right people. The transitioning of units made it difficult. In general, when a unit rotates out, they don't necessarily understand the importance of the data. If there's a way we could match Red and Blue force data in an easy medium, that would be awesome. It is not easily available. Some people have tried to mine it, but there's not much data, you can only get part of it. I had to try to acquire BFT data from the contractor, but contractors don't understand the timeliness of the request. A lot of data are proprietary, but that's [expletive].[19]

This analyst's experience may or may not reflect the experience of all analysts across OIF and OEF, but several other analysts also identified the lack of accurate and complete BFT data as an opportunity for improvement. Some analysts felt this might be the best way for DoD to make a useful investment in data technology.

Summing Up

These studies are examples of how problems supported by quantitative data are much more amenable to analysis using any tool. The flight times, ring-routes, C-130–capable airfields, the AMRs, etc., provided the data needed in the lift study to serve as inputs to the MASHPAT model—and the data allowed for model validation, something critical to analysis. In the surgery team location problem, the nearness, security, and access variables fed a simple linear regression model.

In the lift study, it is clear that the type analysis using the MASHPAT model or any other tool would not likely be conducted in theater. In fact, this study was con-

---

[19] Interview with analyst.

ducted by analysts at the TRAC facilities and Transportation Command facilities at Ft. Lee in Virginia and Scott Air Force Base in Illinois, respectively. In the surgical team location study, just the opposite is true. As long as the data are readily available, this type of problem is definitely suited for analysis in theater.

CHAPTER FIVE

Support to Campaign Assessment Decisions

The previous two chapters described ways in which modeling, simulation, and analysis were generally effective in supporting commanders' key decisions in OIF and OEF. The tangible, often tactical requirements associated with force protection and logistics allowed analysts to identify opportunities to reduce casualties, undermine the insurgents, and improve efficiency by saving time, wear, and cost. Campaign assessment offers few if any opportunities to so clearly link analytic effort with a positive operational or strategic outcome. Interviewees who supported campaign assessment analysis reported the frustrations associated with a poorly defined problem, inadequate data, and a lack of common, validated methods.

Commanders had a hard time articulating their requirements for IW/COIN campaign assessment, and analysts had a particularly difficult time trying to address key decisions for assessment. In IW, commanders and analysts are faced with a problem that has no clear end state, no clear design, data of unknown completeness and unknown quality, no clear way to determine the relevance of any specific variable to an (already unclear) outcome, and no theory that would describe how multivariate quantitative analysis could reveal a path to success or failure.

Analysts expressed particular anxiety over their inability to answer questions with sociocultural elements. When asked to assess corruption in Iraqi ministries, one analyst asked, "How do you account for cultural expectations? Even FM 3-24 talks about 'culturally acceptable' levels of corruption."[1] Analyst anxieties were not restrained to sociocultural factors. When asked to assess ANSF capability, one analyst was frustrated by the opacity of available reports. Did a "partnered" operation "mean you grabbed an ANA [Afghan National Army] solider on your way out of the FOB, or that ANA officers led planning and execution?" These distinctions had important meaning for the analyst's ability to determine how developed ANSF capabilities were, and whether a particular region was ready to be transitioned to ANSF control.[2]

---

[1]   Interview with commander.

[2]   Interview with commander.

This chapter details commanders' concerns and doubts regarding the viability of current approaches to campaign assessment and similar doubts voiced by analysts. It reveals aggressive, often inventive efforts by analysts to address this complex question, but it does not reveal clear connection between these efforts and successful decisions. The absence of that evidence in these anecdotal vignettes does not prove these efforts were fruitless; the process of thinking through problems is usually helpful in some regard. However, previous RAND research shows that more than 300 commanders, analysts, and civilian consumers of campaign assessments were dissatisfied with the current process as articulated in doctrine.[3] Much of that frustration was evident in our interviews and also in the literature review.

Commanders' Decisions in Campaign Assessment

Commanders reported that they used assessments to evaluate progress toward campaign objectives in the operating environment, progress toward execution of the campaign plan by coalition forces, or the effectiveness of coalition concepts of operation. An assessment might incorporate both quantitative and qualitative data, though in practice, they have been closely associated with quantitative methods and metrics in both OIF and OEF.

Some commanders held deep reservations over how well coalition forces truly understood what was going on in their area of operations, and hoped analysts would be able to provide additional insights into either the efficacy of particular tactical actions, such as establishing a Village Stability Operations site in a given district, the meaning of particular events, or broader assessments of how their areas of operations were progressing against the unit's lines of effort.[4] Other commanders distrusted quantitative analysis and preferred to depend on their subordinate commanders' judgment.[5] Several commanders voiced concerns over the validity of the assessment process, noting that the process is "too numbers focused . . . You end up with a stoplight chart, instead of context."[6] Many commanders felt that the metrics identified by the local commander were more relevant than those driven by higher echelons that lacked the context to understand the metrics they asked for.[7] These concerns about context are reinforced by

---

[3]   See Connable, 2012. Note that this observation does not include recent improvements to the ISAF campaign and strategic assessment processes.

[4]   Interview with commander.

[5]   Interview with commander.

[6]   Interview with commander.

[7]   Interview with commander.

the anecdotes of one analyst: "We ended up generating stoplight charts with 14 grada-tions—because that's how many colors were available in PowerPoint."[8]

Global force management for extended campaigns may undermine the trust requisite for a commander to depend solely on subordinate commanders' judgments. Commands regularly deploy with nonorganic maneuver units with which they haven't previously trained.[9] Therefore, while a division commander might want to rely on the assessment of his subordinate brigade commanders to support his division assessment, he might not feel comfortable doing so as new and unfamiliar commanders rotate into theater.

Figure 5.1 depicts key decision points that commanders identified in the inter-views conducted for this report.

Operating Environment

Commanders used assessments to help determine whether they were advancing along their lines of effort. Typical lines of effort assessed include security, host-nation secu-

Figure 5.1
Commanders' Decisions in Campaign Assessment

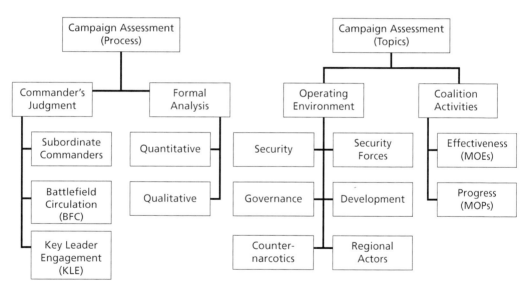

NOTES: MOPs are measures used to determine U.S. unit performance. For an alternate organization of decisions to be assessed, also see Irregular Warfare Methods, Models and Analysis Working Group, "Final Report," scripted brief, Ft. Leavenworth, Kan.: TRADOC Analysis Center, August 18, 2008.
RAND RR382-5.1

---

8   Interview with commander.

9   Interview with commander. Also see Wesley Morgan, "Afghanistan Order of Battle," Washington, D.C.: Institute for the Study of War, June 2012.

rity forces, and governance and economic development.[10] Commanders expressed the greatest uncertainty concerning sociocultural and political factors and events. "Is it a big deal that an elder has come back [from being an expatriate]? Is it a big deal that Badr Corps is fighting with Jaysh al-Mahdi? Frequently we don't have a good baseline to compare today's [area of responsibility] to."[11]

Coalition Activities

Commanders use assessments and associated metrics both to ensure the campaign is being executed according to their vision and to update their beliefs about the efficacy of the current concept of operations. The former is essentially a management function, exploiting measures of progress. The latter activity seeks to validate the current campaign design—or adjust it if necessary—and involves the linkage of coalition activities with measures of effectiveness.

Assessing the effectiveness of coalition efforts implies a causal linkage between coalition efforts and the outcomes observed. Given the complexity of the operating environment, that linkage is not a simple effort. One analyst highlighted that efforts to assess the effectiveness of a given tool becomes problematic when abstracted from its local context, noting that a DoD intelligence organization had assessed the effectiveness of the Commander's Emergency Response Program (CERP) based on what impact its expenditures had on SIGACTs. This approach ignored that CERP expenditure rates might be systematically different during clear, hold, and build phases of an operation; be used differently in each phase of the operation; and so have different effects in each phase of the operation.[12] As the ORSAs in the 2010 MORS conference noted, campaign analysis often contributed to understanding rather than showing causality or correlation.

Vignette: Determining the Impact of the Sons of Iraq on the Campaign

From mid-2006 through late 2007, the Sunni Arab population went through an "awakening" in Iraq. Citizens rebelled against Al Qaida in Iraq by joining local militias, which were typically referred to as either Sons of Iraq or Concerned Local Citizen (SoI/CLC) groups. Theater-level commanders in Iraq had to determine whether they should back these local militias; doing so would entail significant risk. If these militias failed, the Americans could be accused of backing local warlords. Arms provided to

---

[10] Interviews with commanders; DoD, *Report on Progress Toward Security and Stability in Afghanistan: Report to Congress in Accordance with the 2008 National Defense Authorization Act (Section 1230, Public Law 110-181)*, Washington, D.C., October 2011.

[11] Interview with commander.

[12] Interview with analyst.

militiamen might have wound up in the hands of insurgents. Commanders and analysts in Afghanistan faced similar issues when determining whether to back the groups that evolved into the Afghan Local Police. In Iraq, members of one senior analytic staff took it upon themselves to support this decision with forecasting analysis.

Analysis

Analysts at the theater level saw the problem existed, and that the commander was faced with making a key decision in the absence of sufficient analysis. "Without . . . a direct tasking, our little shop kind of figured out that we needed to do something to support the decision or at least inform the commanding general in his discussion with [senior political leaders]."[13] This self-initiated analysis was designed to determine the downstream effect of backing the SoI/CLC groups on the campaign. What results would these groups have in terms of campaign progress? What are the potential costs and risks involved? The analysts compared investment in each group in each specific area with environmental feedback in that area. This was a predictive data correlation analysis: If we invest X amount of money in a specific group in a specific village, how does this translate into a reduction in IEDs found and cleared, detonated, or emplaced? What is the actual dollar figure per successful IED find?

Data

This analysis relied on the use of SIGACT IED data and casualty data compared with money and equipment investments in various SoI/CLC programs. As ORSAs at the 2010 MORS conference, and in the CAA Deployed Analysts History, noted, SIGACT data were considered "dirty" in that they were rife with errors in category, type, location, and other critical factors for each attack.[14] ORSAs and other analysts supported the development and maintenance of the SIGACT III database, which reflected efforts by ORSAs to clean the data. This would involve correcting acronyms, making location names more accurate, and performing other clarifying tasks. Both the SIGACT and SIGACT III databases evolved over time, with categories being added or removed, or definitions of key variables being changed.[15] Therefore, analysts would also have to periodically conduct retroactive adjustments of datasets and reports to account for these changes.

Results

"[The commanding general took our results] and walked into the video teleconference with the [senior policymakers] and that was the discussion."[16] No empirical validation

---

[13] Interview with analyst.

[14] CAA, 2012.

[15] CAA, 2012, pp. 80, 87, 94 and 95.

[16] Interview with analyst.

of these results is available. They were probably generally informative rather than deterministic. As with most other campaign analyses (or assessments) in COIN, trying to prove the real value of the analytic results is frustrating. In this case, as in many others described in this report, the results of the analysis may have been less important than the process of thinking through the potential variables at play and identifying likely correlations. As long as the commander and the policymaker did not take the analytic results at face value—data are generally poor and no forecasting analysis is ever truly accurate—then there appears to be good use for this kind of analytic exercise.

Issues

Data for U.S. casualties are generally highly accurate, but data for Iraqi military and civilian casualties are less so. While there are often discrepancies, gaps, and inaccuracies in the SIGACT databases in both Iraq and Afghanistan, IED data tend to be more accurate than other types of attack data because of the many crosscutting programs associated with IED reporting and analysis. However, it is not clear how strong a correlation could be drawn between dollars invested in the SoI/CLC programs and an actual reduction in IED effectiveness; many other variables could also affect a change in IED reporting. As with many such analyses, isolating the variables to create meaningful outputs is difficult to accomplish and defend.

SIGACT data are never complete in that they cannot and do not represent all of the insurgent-related violent acts that occur across a theater like Iraq or Afghanistan on a given day. One could assume that violent acts against coalition troops are more likely to be reported, and reported accurately, than violent acts against local security forces. While there is no empirical data to show how various issues affect SIGACT data quality, it would be safe to say that the less mature and organized the local security force, the less likely they are to report attacks completely or accurately. For example, a highly skilled Iraqi special forces unit with a coalition adviser team would be far more likely to report attacks completely and accurately than a newly formed militia unit that might not even have a formal means of communications. The very act of turning over battlespace to Iraqi militias could skew the IED data since the accuracy and completeness of reporting in those areas would necessarily decline in the absence of persistent coalition presence.

It is not clear how the output from this analysis was incorporated into a broader campaign assessment, if at all. In many cases, these kinds of comparative data analyses were produced as stand-alone products and used to inform rather than shape campaign assessment. This report appears to have been tailored for the specific purpose of validating SoI/CLC. However, it almost certainly helped the analysts and the commander think about variables and also to identify gaps in collection and knowledge.

Vignette: Automating Campaign Assessment

A number of interviewees described the challenges associated with managing and trying to interpret the massive data feeds that were piped in to the staffs on a daily basis. Computer-aided systematization or automation was viewed as a necessary step to help process these data for campaign assessment. In this case, a theater-level assessment staff systematized the data capture process and then automated the generation of trend reporting.

Analysis

The theater-level staff that undertook this data process did so on their own initiative and using locally built tools: "We were home-brew for the most part."[17] In other words, the tools were all developed in the field using a common Microsoft Office tool kit and macro scripts. The assessment analysts captured data in a Microsoft Access database, channeling input from a Java link generator that they provided to each subordinate command. These commands would receive the link and use it to input their monthly data, including such information as numbers of operations of various types, and unit readiness. The process was designed to make this as easy as possible for subordinate units.

Analysts in this case referred to these quantitative data as "objective metrics" in that they viewed them as objective rather than subjective.[18] They also added coded "subjective" metrics into the database, including commanders' assessments of IO effectiveness, and unit ratings for each province based on a 1–100 scale. For example, a subordinate commander might "rate" a province it controlled as a 73 out of 100 based on the staff and commander's internal assessment of the province. This number would then be inducted into the Access database and generated in a report.[19]

Once the data were collected and other data were pulled in from theater-level databases, the analysts would use scripts to transfer the data into Microsoft Excel spreadsheets. From there they would generate charts, which they would advance from month to month to develop longitudinal plots. Once the charts were built in Excel, the analysts would activate another set of scripts to generate Microsoft PowerPoint briefs. These briefs would display the aggregated data in a "traffic light" metrics report: negative results would be displayed in red, middling results in yellow, and positive results in green. This entire process was eventually automated: "Assuming you had already collected the data, you could just hit play and it would generate your slides for you."[20]

---

[17] Interview with analyst.

[18] Interview with analyst.

[19] It was not clear whether this was an ordinal or ratio scale. It was most likely an ordinal scale reflecting the subjective interpretation of the commanders and staffs without having any quantitative meaning in the real world; "73" would not be correlated to an actual set of calculable indicators.

[20] Interview with analyst.

Microsoft Office played a central role in this process because it was available and useful: "I emphasized just the Office suite—it's on every desk. It's a fact of life, in terms of what people expect out of you [for] products and in terms of data." Also, "Stats package is on pretty much every desk . . . but a stats background is essential [to use this]. ArcGIS was very cool because that took things from the database and put them on maps . . . Geospatial is very cool but it's very specialized."[21]

Data

Since this process was designed to capture and plot data across all lines of operation, nearly any quantitative data type was viable. The analysts adjusted the type of data they would pull based on their interpretation of current operations and current requirements. They would try to continue to collect similar data over time to develop a basis for longitudinal analysis when this was possible. Data sources were readily available, but database control was a noted challenge. "We'd pull [data] in from different sources. Operational databases were big . . . But after 72 hours nobody cared about whether your particular database was right any more. So, we had one of our operations guys whose job was to create a derived operational database and . . . correct the data."[22] As with all data in the IW environment, there was no way to determine the actual accuracy and completeness of each data set as it related to ground truth, at least not from the unit headquarters.

Results

The analyst stated, "We got good feedback." But in some cases the consumers questioned the output or process: "Someone must have thought a metric was fishy and have confusion—you know, 'how did you arrive at that metric?'"[23] This automated process tended to support more detailed data analysis. "Typically [commanders and staffs would use this to] drill downs into the metrics. You know, 'tell us what you know about a particular kind of problem,' or 'give us an example of how . . . an infrastructure issue changed over time.' They [consumers] would see one of our operational metrics up on a slide and ask, 'well, what changed from last month? Can you tell us why?'"[24] The interviewee did not identify any specific decisions supported by this tool, nor provide any vignettes that would describe how the tool supported a positive change in the campaign. It seemed that this was primarily an informative tool rather than something used for direct decision support.

---

[21] Interview with analyst.

[22] Interview with analyst.

[23] Interview with analyst.

[24] Interview with analyst.

Issues

While Office was prolific and in common use, the ad hoc Office scripts were not without flaws. "The thing was, they're very delicate. They break all the time."[25] This problem was compounded by the rotation of analysts in and out of theater. Often the analyst who wrote a specific script would have rotated home when that script failed, and the new analyst would have significant trouble fixing the script. "Once it breaks, it isn't useful any more. Then you'd go back to nothing. So that's what kneecapped the tool chain we were using. It's all about being able to get the tool and maintain it."[26]

Tools were very useful to this analyst, but they needed constant support by experts in specific technical specialties who were in high demand. "We had another guy in our shop who was the 'macro guru,' and guys from other sections could come—'hey man . . . please fix this for me.'"[27] This requirement for expertise tied back to reflections on analyst training. "They had a crash course on how . . . to become more efficient with Microsoft Office . . . 'This is ArcGIS, these are some simple database queries you can use.' It's great because it's a week where you're familiarized with the selection that you have while in theater."[28]

While these tools automated the collation and display of various data, it is not clear from these interviews that the automated tools provided any kind of critical thinking or in-depth analysis in support of commanders' decisions. Because the analysts were tied up with data and tool management, it is not clear that they had any time to read, understand, or provide value-added insight into any of the data or the time-series charts. The analyst describing this process seemed, through no personal fault, wholly engrossed with the process itself and less focused on the ways in which it may have supported decisions.[29] This approach may have been necessary to provide commanders with the data they demanded; ORSAs are not completely free to choose their approaches and methods. But this particular type of process did not align with the ten-step, logical OR progression identified in Chapter One of this report.

Nor is it clear how the metrics for this automated process were selected. Based on the interview, it appeared that the analysts selected the metrics based on data that were (a) available and (b) relevant (in the analysts' view) to the commander's priorities. This seems reasonable in the absence of other, clearer guidance, and it is practical in some sense if the commanders are thoughtful and conscientious in updating their Commander's Critical Information Requirements, a list of priority information needed to support key decisions. However, the absence of a clear connection between end-state

---

[25] Interview with analyst.

[26] Interview with analyst.

[27] Interview with analyst.

[28] Interview with analyst.

[29] This is in part a function of the questions asked, but it is also reflected in the emphasis found in the interview responses.

objectives, a path to end state, and well-reasoned indicators begs the question of over-all utility for data automation: While an automated process is easier to use and read, it is not altogether obvious that it provides the commander with the information he needs to understand the flow of the campaign. It would seem that availability of data should not be conflated with the relevance of those data to the complex question at hand. Many ORSAs have recognized this problem and, as we showed in Chapter Two, they have identified the need for deployed and reachback analysts to help commanders identify and validate MOEs, MOPs, and indicators. It is worth considering whether commanders could have better employed their analysts in support of general planning rather than in managing data or attempting to quantify and qualify individual variables.

Vignette: A View of Campaign Assessment from the Top

In this case, the interviewee was a theater commander in a counterinsurgency campaign. The commander provided insights into his efforts to try to understand the ebb and flow of the campaign. He was particularly concerned with determining how the population was responding to coalition efforts, and how the enemy was able to shape the population's attitudes toward the coalition. He described his experiences with reachback support, which he found critical to his overall efforts. He was less sanguine about the use of tools and simulations to support his decisions.

Analysis

The theater commander thought it was necessary to tap into nonorganic analytic support to provide insight into population dynamics. "At the strategic level, you take advantage of academics . . . and think tanks."[30] He believed that solicitations from SMEs were pivotal for campaign assessment, and anthropology was particularly useful. Structured techniques also contributed to assessment and decision support, but to a lesser extent. While the commander did not rely heavily on a single or set of structured methods, he did see value in longitudinal analyses of polling data. He relied on these simple longitudinal analyses, presented in charts, to shape his personal assessment. These polling analyses were done primarily through reachback support, and in conjunction with university partners and academic researchers. He acknowledged that this was an ad hoc approach, and that he was trying to find analytic approaches that worked rather than analyses that were proscribed in doctrine or organizational protocol.

Other analyses focused on political durability based on information derived from sociocultural collections in theater. The commander wanted to know whether the provincial and local officials would survive in office, if they would be replaced, or if they

---

[30] Interview with theater commander.

would even support the government. He asked his reachback analysts to provide him with an assessment of durability based on bottom-up sociocultural data compared with data on rule of law, economics, governance, and other factors. This was, for all intents and purposes, intelligence analysis, but it served as a fine-tuned and tailored feed into the commander's campaign assessment.

Data

The commander relied on multiple data streams across all lines of operation to inform his campaign assessments. Basically, if it was available and he felt it was useful, he put it to use. He pressed his staff and his collection assets to obtain more, and more granular, sociocultural data so he could learn more about the population. He felt it necessary to have people with experience in social sciences out in the field to get those data for him.

Results

Analytic output provided this commander with more of an impression than a clear understanding of the issues and trends. However, this impression supported his decisionmaking. He emphasized that commanders needed to seek out multiple lines of analysis using multiple methods, in essence triangulating to obtain a better understanding of ground truth.

Issues

There are inherent dangers in every analytic approach at the theater level due to issues of data aggregation, but also the compounded complexity of the problem: the theater commander must take all aspects of the campaign and all layers of activity into account. The commander stated that the structured techniques like those suggested in joint doctrine are ineffective because the situation in theater is so dynamic: "Mechanistic approaches don't work; there's no cookbook" for assessment.[31] In general, the commander was suspicious of structured analyses, modeling, and simulation. He felt there was an assumption among analysts that the most complex and intricate analytic and M&S tools were somehow better and more useful, but "that is exactly not right." He believed that some analysts were overly enamored with their toolkits, to the point that they lost sight of reality. "There should be a logo when you do this work: No Hubris."[32]

---

[31] Interview with theater commander.

[32] Interview with theater commander.

Vignette: ISAF Joint Command District Assessments

In late 2009 and early 2010, the IJC created and implemented an assessment process to determine campaign progress in the "Key Terrain Districts" (KTDs) of Afghanistan. Because there were approximately 400 districts in Afghanistan, and many of those were of limited relevance or generated limited feedback, IJC focused their efforts on KTDs so they could produce assessments that were relevant to the IJC commander and, ultimately to the ISAF transition plan.[33] This winnowing process also recognized the limits of their assessment capabilities: They initially had a very small staff of analysts in their Information Dominance Center (IDC) dedicated to the task.[34] These analysts were a mix of ORSAs and coalition combat arms officers (e.g., an Italian mortar company commander) pressed into service as analysts. They were tasked with assessing approximately one-third of the KTDs every six weeks in collaboration with the regional command (RC) staffs. In this vignette, the interviewee was an experienced midlevel analyst assigned to the IDC to support the District Assessment process.

Analysis

"The ultimate question was: How are we doing in that district [across] the lines of operation?"[35] The District Assessment represented an effort to provide a district-by-district, color-coded assessment that would indicate success in developing security, governance, the economy, and also rule of law. Figure 5.2 depicts an early iteration of the District Assessment report. The KTDs are coded by color, with gray representing the KTDs not assessed during that period.

At first the IJC team did not conduct any analysis. They took the input from the RC and formatted it for the report. According to the interviewee, this approach reflected the IJC Commander's preference to give considerable leeway for assessment to his subordinate commanders. He trusted their instincts, and he "just wanted to hear what they're saying."[36] The analysts worked with the IJC commander and staff to add what they believed was analytic rigor to the process. They worked from the assumption that the RC staffs did not have the time or the assets to conduct thorough analyses, and that the fluctuations in assessment during relief-in-place/transfers of authority would lead to significant ripples in the longitudinal assessment. Essentially, every new RC commander coming in would reset the color code based on his personal viewpoint. This made it nearly impossible to track progress over time. This kind of fluctuation

---

[33] There is ongoing debate over the actual number of districts in Afghanistan since the creation of the Marjah District in the aftermath of Operation Moshtarek.

[34] In late 2011 and early 2012, IJC grew this staff to help address the pressing demands for information and also for transition assessment.

[35] Interview with midlevel analyst.

[36] Interview with midlevel analyst.

Figure 5.2
District Assessment Report Map

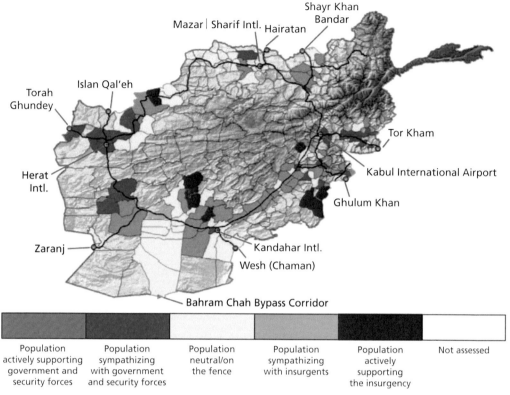

SOURCE: Sang Am Sok, Center for Army Analysis, "Assessment Doctrine," briefing presented at the Allied Information Sharing Strategy Support to ISAF Population Metrics and Data Conference, Brunssum, Netherlands, September 1, 2010, p. 36.
RAND *RR382-5.2*

is reflected in Figure 5.3, which portrays exaggerated change to help visualization. It depicts increasing confidence by commanders over time, and then a drop in confidence when a new commander appears:

There was no time to build or apply models to support the District Assessment process. "The data situation was so intractable that even if we had tools and models, what we put into them would not have necessarily led to [believable] results."[37] Ultimately, the analysis at the IDC during this period was conducted by brainstorming. "We took polling data, focus groups, qualitative research, etc., and we sat down and quite painfully said, 'OK, what do we have?'"[38] Analysts attempted to find correlation between data streams and produced a color code based on their professional judgment.

---

[37] Interview with midlevel analyst.

[38] Interview with midlevel analyst.

Figure 5.3
Impact of Relief-in-Place/Transfer of Authority on Assessment

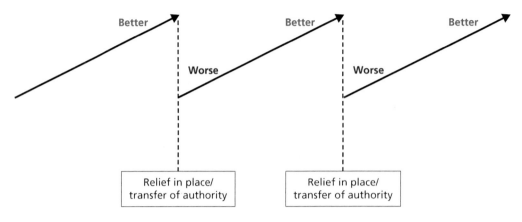

SOURCE: Connable, 2012, p. 224.
RAND *RR382-5.3*

Data

The assessment analysts did not have the authority to direct data collection ("we weren't *collecting* data"), but they could use data from large databases like the Combined Information Network Data Exchange, or request existing data from RCs. At first, they did not use any data; they accepted the RC-level inputs. Later, the assessment staff began to draw its own data for what it called "objective" analysis to check the RC assessments. The team built a set of more than 100 quantitative metrics, deriving the list from multiple SME inputs. This list bounded their analyses and also helped them to effectively channel their data sources. While the list is not available to the public, it included metrics and associated data sources across all the lines of operation. Essentially, any data that were available and relevant to the metric might be used for assessment.[39]

Problems with data quality, consistency, and availability were endemic, particularly in the early stages of the surge in Afghanistan. "There was no consistent reporting . . . there was no standard concept" for inputting data at the tactical level.[40] The analyst observed that this was true for SIGACTs, but particularly for ad hoc inputs like Key Leader Engagement reports that, over time, came to be seen as a critical reporting stream for the analysis of governance, economics, and rule of law. "There's no standard concept of what a key leader is, what an engagement is, and what should be reported."[41] The analysts also had no way of knowing if the data and even the basic information

---

[39] This assessment of the IJC IDC process is derived from a series of interviews conducted by RAND in 2010, 2011, and 2012 as part of ongoing research on the Afghanistan assessment process.

[40] Interview with midlevel analyst.

[41] Interview with midlevel analyst.

like the date and location of the report were accurate. The analyst believed that this kind of inconsistency in data collection undermined the campaign: "By not collecting this data well and not collecting it with any consistency, [we] forfeited the ability to learn from past mistakes."[42]

Results

While the analyst did not point to any specific decisions that this campaign assessment process supported, the results in this case—as in similar cases—may have been in the process itself. At first, there was no assessment method applied at the IJC level, but the analysts quickly imposed some structure and started experimenting with various approaches. These were ad hoc and unfortunately undermined by the lack of relevant or available data, but they did provide alternative input to the District Assessment.

The value and effectiveness of the District Assessment process is hotly contested in the circle of SMEs familiar with campaign assessment, and the methods that have been employed and altered over time are controversial.[43] Over time, however, having a team of analysts think through the problem of aggregated campaign assessment at a major command, in theater, has provided the entire community of analysts with an opportunity to witness the challenges of assessing an IW/COIN campaign. Insight provided by these assessments have influenced changes in the campaign assessment process at the ISAF and at the RC level, as well as influenced input into the early 2012 efforts to rewrite FM 3-24.[44] Therefore, even if the output of the process was not clearly effective, the process itself was useful and informative.

Issues

One could argue that the imposition of structure introduced artificiality to the assessment process, and that the IDC analyses diverted from the value and integrity of the RC and IJC commanders' professional judgments. One could just as easily argue, as many analysts have, that it is necessary to balance subjective assessments of commanders with objective analysis of (primarily) quantitative data. In the context of the earliest days of the District Assessment process (2010 to early 2011), this predilection was reinforced in contemporaneous doctrine, which stated: "Quantitative indicators prove less biased than qualitative indicators. In general, numbers based on observations are

---

[42] Interview with midlevel analyst.

[43] For example, see Jonathan Schroden, "Why Operations Assessments Fail: It's Not Just the Metrics," *Naval War College Review*, Autumn 2011; Downes-Martin, 2011. Also see Connable, 2012, and Claflin, Sanders, and Boylan, 2010.

[44] These observations are based on RAND's participation in, and observation of, the assessment processes at all levels of command in Afghanistan from 2009–2012, as well a participation in the 2012 Field Manual 3-24 rewrite conference.

impartial."[45] This bears more broadly on the general concept of applying modeling, simulation, and structured analytic techniques to key commanders' decisions. It is worth considering whether the assessment of trusted field commanders stands alone, whether the commander's assessment would benefit from rigorous input, or whether quantitative analyses are in fact more objective—and therefore, one must assume, more valuable for decision support—than commanders' assessments or some form of qualitative analysis.

Vignette: ISAF Campaign Assessment

In 2012, the ISAF Commanding General, John Allen, directed the ISAF AAG to create a new theater-level assessment process. This new process splits the assessment into a campaign and a strategic assessment: The former addresses progress toward executing the ISAF campaign plan while the latter addresses progress toward NATO strategic plans for the Afghanistan-Pakistan region. Campaign assessments are assigned to subordinate functional commands such as IJC, which provide a combination of a narrative report with a subjective numerical judgment of progress across several lines of operation such as security. Commanders may include contextualized quantitative data in their narrative assessments, but they are not required to do so.

This process places a premium on commanders' inputs, and eschews a purely quantitative approach to assessment; the differentiation between objective and subjective analysis has been blurred, or perhaps made less relevant. The strategic assessment is an ISAF staff process. Each staff section responds to a series of strategic questions with a long-form narrative report. These reports typically include both quantitative and qualitative data, but the format requires that these data be presented in context.[46] The commanding general retains responsibility for all assessments that leave ISAF for higher echelons of command. Therefore, the commander's *coup d'oeil*, or eye for the situation, is the final arbitrating point.[47]

Analysis

The need for a new campaign assessment process came about due to the over-reliance on purely quantitative assessments that built on hundreds of effectiveness and performance measures, or quantified subjective assessments in which judgments were simply translated to numbers or color codes. The quantitative assessment process dates to late 2009 when ISAF's newly formed Afghan Assessment Group proposed a campaign

---

[45] Headquarters, U.S. Army, *The Operations Process*, Field Manual 5-0, Washington, D.C.: March 2010, pp. 6–8.

[46] These observations are based on previous and ongoing RAND research in support of the ISAF assessment process. See Connable, 2012.

[47] The literal translation from French is akin to "stroke of the eye."

assessment model that generally followed guidance in U.S. Army and U.S. joint doctrine.[48] This comprehensive assessment process included approximately 40 pages of effectiveness and performance measures, and demanded the collection of several hundred metrics, or indicators, like the daily number of violent incidents (SIGACTs). The Commanding General of ISAF reportedly dismissed this carefully developed, doctrinally sound process as too complex and unrealistic for Afghanistan. The District Assessment process was developed to fill the gap left in 2009, and ISAF subsequently struggled to find a formula that would both be realistic and meet consumer demands.[49]

Between 2009 and 2012, staff officers at ISAF experimented with a range of quantitative and narrative assessment methods, often differentiating between the two rather than seeing an opportunity to integrate methods or formats. As they gained experience with the available data and with the complexity of the environment, they began to recognize the inapplicability of U.S. and NATO conventional assessment doctrine in a COIN environment. While large quantitative models might help them conceptualize the IW problem, these models could not account for all of the variables, data issues, and causal linkages presented in the real world. Figure 5.4 depicts a model of an IW/COIN theater-level campaign that in many ways reflects both standard U.S. assessment doctrine and also the proposed 2009 quantitative approach to assessment in Afghanistan. This model depends on more than 400 interdependent ontologies, cross-referencing DIME actions with actors (individuals, groups) and environmental factors to show causal impact. Each red, green, yellow, and white rectangle represents an ontology, or variable.[50]

This model might be useful to support the development of a computer-aided simulation or noncomputer-aided wargame; in fact, this model directly supported development of the TRAC Irregular Warfare Tactical Wargame (IWTWG).[51] Simply thinking through all of the possible variables, permutations, and interactions depicted in this model could be enormously helpful to COIN and IW planning and training. Yet it would seem unreasonable to assume that any model of this complexity could hope to discern causal relationships using shifting, incomplete, and often inaccurate real-world data. If applied to actual ongoing operations, this approach—as reflected in this model and in the 2009 ISAF assessment process—would appear to far exceed

---

[48] At this point in time, NATO campaign assessment doctrine was limited and not necessarily applicable to IW assessment.

[49] See Connable, 2012, for a discussion of this process. This observation is derived from interviews conducted for *Embracing the Fog of War: Assessment and Metrics in Counterinsurgency.*

[50] The briefing indicates that each independent variable is described as an ontology because each was carefully analyzed and vetted with SMEs to determine likely attributes. For the purposes of this report, they can be equated with variables.

[51] Hartley, Lacey, and Works, 2011.

Figure 5.4
A Model of an IW/COIN Campaign at the Theater Level

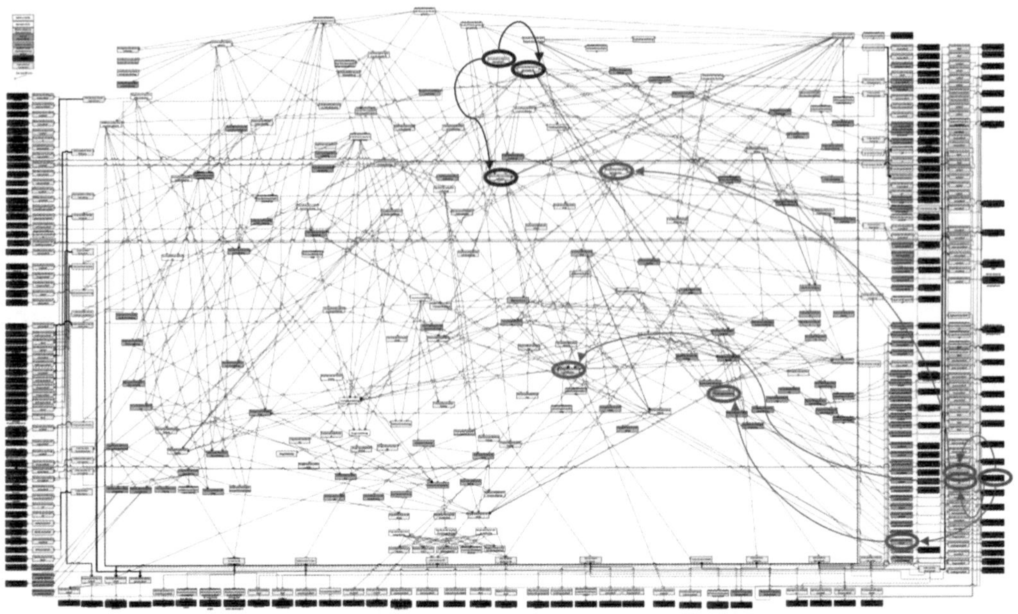

SOURCE: Dean S. Hartley, Lee Lacey, and Paul Works, "IW Ontologies," briefing, INFORMS National Meeting, Charlotte, N.C., November, 2011.
RAND RR382-5.4

the capacity of modeling to provide realistic input to decisionmaking.[52] ISAF officially reached this conclusion in 2012.

Data

The primary data source to support the theater-level assessment process is a set of narrative responses to a series of strategic questions with a long-form narrative report, generated by each command staff element, that are subjective but derived from command field reports. The narrative reports can also include quantitative data, but staff elements are required to place the quantitative data in context.

Results

This new process has had some growing pains, and it is no panacea. Campaign assessment for IW/COIN will always be a messy process, and both commanders and analysts will be left somewhat dissatisfied by both the process and the results. A large portion of the ISAF staff continues to work with out-of-context quantitative data, providing charts to support campaign-level decisions. However, an assessment that includes a

---

[52] There are no indications that this specific model was used to support real-world operations.

subjective evaluation supported by sound field reporting and presented in narrative form has proven to be an acceptable method.

Issues

This new approach to campaign assessment represents a dramatic departure from both existing doctrine and recent practice. ISAF undertook this deviation only after attempting to implement existing doctrine in various forms and iterations, each of which left senior commanders dissatisfied. However, ISAF's new approach does reflect an effort to address a heretofore unresolved, inherent conflict in campaign assessment for IW: The approaches, methods, and tools that have proven useful for tangible problems, such as force protection and logistics, have not proven useful for real-world (as opposed to simulated) campaign assessment.

Summing Up

It is unlikely that COIN or IW campaigns will ever have clearly defined end state objectives or that they will provide analysts and commanders with accurate or uniformly relevant data. The nature of IW, and particularly of large-scale COIN, means that challenges, objectives, timelines, and requirements will vary from place to place within a theater, so the kind of uniform, centralized metrics and approach used for conventional assessment is unlikely to work for IW campaign or strategic assessment. Experience in Iraq and Afghanistan has provided the necessary building blocks to develop new analytic methods and doctrine: shortfalls, a series of what arguably were failures, and at least one major adaptation that might have generalizable and enduring applicability. The ISAF campaign and strategic assessment process that currently stands as a major diversion from doctrine should be considered as a model for new IW assessment doctrine, at least for COIN.

CHAPTER SIX

Support to Force Structuring

We have noted that IW campaigns tend to be long and ill-structured. They require varying amounts of resources over time, and it is rarely clear at any one time how many people, how much equipment, and how much money is necessary to achieve victory. As campaigns drag on, policymakers are faced with not only minimizing costs, but also sustaining popular support. Vietnam, OIF, and OEF showed that sustaining high concentrations of troops and spending large sums of taxpayer money becomes increasingly difficult over time, particularly since weaknesses in campaign assessment methodology make it so difficult to convey meaningful progress toward a clear end state. Policymakers demand that commanders tell them how many troops are required to win, for how long, and why. Senior commanders are then drawn into a complex policy debate. These commanders are responsible for providing policymakers with a clear rationale for their force structuring requirements, but the analytic community has not yet provided them with a methodology that provides a clear rationale for IW force-structuring requirements.

In general, commanders and analysts have taken one or more of the following approaches to try to determine strategic force requirements—the most critical and widely debated aspect of IW force structuring—for COIN. This subject is explored in greater depth in a range of professional articles, briefings, and in RAND's *How Insurgencies End*.[1] This RAND report provides a fuller examination of the troop ratio and troop density calculations.

Troop Ratio Calculations: Also called force ratio calculation, this process establishes a troop requirement by first determining the number of enemy combatants and then determining how many U.S. or coalition forces will be needed to defeat these insurgents. Suggested ratios vary, but this approach is undermined by the proven inability to accurately count enemy forces—particularly insurgents—in IW. Most enemy force calculations are actually educated guesses based on very shaky data and often equally shaky assumptions. Further, if IW and COIN specifically are population-centric operations, then it is not clear that an enemy-centric calculation is warranted.

---

[1] Ben Connable and Martin C. Libicki, *How Insurgencies End*, Santa Monica, Calif.: RAND Corporation, MG-965-MCIA, 2010.

Troop Density Calculations: Also called force density calculation, this is a process of establishing a per capita ratio of friendly forces to the civilian population. As with troop ratio calculations, the suggested ratio varies but is typically rated against increments of 1,000 civilians. For example, FM 3-24 suggests a density of 20 to 25 friendly forces for every 1,000 civilians.[2] However, accurate population surveys are rarely available in IW environments, and population counts are presented as rough estimates even when derived from sophisticated technical means. And as with all other approaches, this one is based on a series of disputed and conflicting historical case studies. Since it is a logical fallacy to use historical cases as a definitive predictor of future behavior, and since there tends to be a high degree of variation among individual IW cases, this approach should not provide the basis for a firm quantitative estimate.

Troop to Task Calculations: This approach is most often used for more tactical applications, but could also be used to help understand strategic requirements. It begins with task requirements—e.g., how many cities must be secured, how many operations must be conducted—and then assigns troops to each task based on historical precedence. But determining specific task requirements over the course of an IW campaign is a nearly impossible challenge, particularly since end-state goals and intermediate objectives change so often and so unexpectedly.

As this report went to publication, there were no approaches to determining strategic force requirements that were considered generally sound and effective by policymakers, commanders, or the analytic community. While analysts have been able to show how forces allocated to a combat theater might be best employed, they are less able to show how many forces should be employed over time. This leaves a critical gap in IW decisionmaking.

The following section addresses some of these issues in greater detail, and also describes several other aspects of force structuring from a commander's perspective. The next section provides a limited number of force-structuring vignettes drawn from our interviews.

Commanders' Decisions in Force Structuring

Commanders' force-structuring decisions ranged from developing a campaign design to defeat the Taliban at the theater level to allocating UASs to track insurgents.[3] For IW, forces are likely to include civilians and contractors. Commanders interviewed for this report described force structuring as part of a four-part process. We do not address each of these in the vignettes, but it is useful to place these key decisions in a broader context. Commanders stated that in an ideal world they would:

---

[2]   Headquarters, U.S. Army, 2006d.

[3]   Interview with theater commander; interview with analyst.

1. begin the IW deployment process with campaign design
2. design their force to meet campaign objectives
3. deploy that force in theater
4. employ that force to achieve campaign and strategic objectives over time.

Commanders, in conjunction with defense officials and policymakers, periodically revisit each of these four steps in order to adjust to changing strategies, environmental issues, and threat actions. Therefore, force structuring is a dynamic, ongoing cycle.

Campaign Design

Campaign design is perhaps the weightiest decision commanders make in IW, possibly across the spectrum of operations,[4] as it shapes the subsequent decisions of force structuring (Figure 6.1). In IW, campaign design typically includes crafting an interlacing range of efforts across security, governance, and economic development lines of operation. Because IW is a warfighting function and typically fought over long periods of time, it is therefore subject to the dynamic fluctuations of war. Campaign design represents the plan that, according to a widely referenced cliché, "does not survive first contact with the enemy." One theater commander noted that he made a deliberate decision to shift from his predecessor's focus on counterterrorism to population-centric COIN. A natural outgrowth of this decision was a shift away from employing forces through raids from large bases, instead making units responsible for the stability of specific geographic spaces and populations.[5] The shift in focus from the enemy to the population entails the integration of both kinetic and nonkinetic activities into campaign planning, one of the key elements distinguishing campaign planning in IW from conventional combat operations. The shift from counterterrorism to COIN also has direct implications for force design, including resourcing and composition.[6] Pouring additional resources into a campaign that is progressing from a flawed campaign design may actually undermine progress toward campaign objectives, or at the very least squander lives, materiel, and time.[7] As one participant in McChrystal's strategic assessment of Afghanistan noted, "Campaign design at the theater level was the critical

---

[4] Irregular Warfare Methods, Models and Analysis Working Group, 2008.

[5] Interview with theater commander.

[6] James T. Quinlivan, "Force Requirements in Stability Operations," *Parameters*, Winter 1995; David W. Barno and Andrew M. Exum, *Responsible Transition: Securing U.S. Interests in Afghanistan Beyond 2014*, Center for New American Security, December 7, 2010.

[7] Stephen Biddle, *Military Power: Explaining Victory and Defeat in Modern Battle*, Princeton, N.J.: Princeton University Press, 2004; Gian P. Gentile, "A Strategy of Tactics: Population-centric COIN and the Army," *Parameters*, Autumn 2009.

Figure 6.1
Commanders' Decisions in Campaign Design

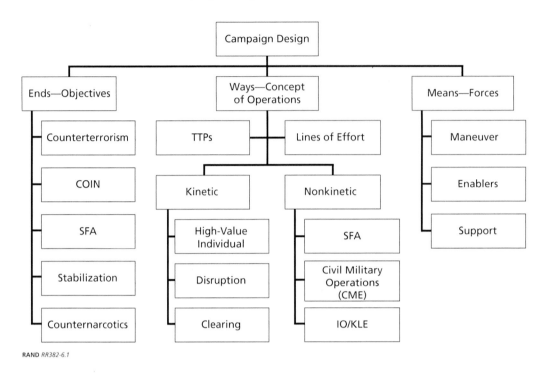

challenge of the war in Afghanistan. If we got that wrong, tactical victories couldn't be meaningful."[8]

There are necessarily only a very few commanders experienced in conducting IW campaign design at the theater level. Even within this small group, attitudes toward analytic support for campaign design vary widely, ranging from disinterest to deep engagement with high-level but ad hoc strategic assessment teams.[9] In Chapter Two we identify several wargames that can support IW campaign design, but there appear to be no widely accepted analytical tools to help explore core questions of IW campaign design—such as analyzing the cost, benefit, and risk of a counterterrorism versus COIN strategy—within a common framework.[10] One theater-level staff director recommended the development of M&S tools to allow commanders to explore a larger

---

[8]   Interview with senior analyst.

[9]   While we interviewed only one of the general officers who served as a theater commander in OIF and OEF, it is possible to obtain a reasonably representative sample of theater commander positions through a review of published public material, speeches, press releases, and existing interviews. Interview with theater commander; Stanley A. McChrystal, *COMISAF's Initial Assessment (Unclassified)*, August 30, 2009. Campaign design should here be distinguished from force design (discussed below), for which theater commanders have received extensive support from the Center for Army Analysis (interview with senior analyst).

[10]   Interview with senior analyst and a review of existing tools.

number of COA than currently manageable.[11] Regardless of the feasibility of that particular recommendation, it clearly points to the desire by senior officers for the ability to carefully examine how robust alternate campaign designs are against different assumptions about the operating environment, and the desire to give commanders more options.

Force Design

Many commanders placed considerable focus on force-design issues, an emphasis we see reflected in the literature on OIF and OEF.[12] Even for conventional combat operations, many elements of the U.S. military organize tasks based on mission. Force-design issues in IW are still fluid in practice, including force-sizing from the theater level to the tactical level, the organization of staff, identification of maneuver units and enablers, and establishing requirements for coalition and host-nation forces. Decisions involved judgments of both the quantity and composition (including training) of required capabilities. Figure 6.2 identifies commanders' decision issues for force design as identified in the interviews and the literature.

Theater commanders can influence the design of the forces provided to them through direct engagement with the force providers and through their Request for Forces. Toward the end of OIF (and OND), the United States deployed brigades organized as Advise and Assistance Brigades, optimized for Security Force Assistance activities in support of the Iraqi Security Forces.[13] During OEF, General David H. Petraeus requested two general-purpose force battalions be deployed to support Village Stability Operations. He then directed their reorganization into largely squad-sized units deployed across the entire theater, transforming the battalion headquarters into a coordination center. These task-organized units assumed responsibility for supporting governance and economic development lines of effort, a dramatic departure from the "battle space owner" mission that infantry battalion headquarters are typically tasked with.

Commanders also influence the capabilities available to them through the operational needs statement process, a tool that allows them to request nondoctrinal, nonorganic capabilities.[14] One theater-level staff director noted, "In June 2009, there was one aerostat [fixed position aerial observation platform] in Afghanistan. There were 75 by the time I left. Programs of Record are aligned with the FYDP [Future Years Defense Plan]. Combat can't wait for the FYDP. So units are given money from battalion on

---

[11] Interview with former theater staff director.

[12] Interview with former theater commander; interview with former staff director; Irregular Warfare Methods, Models and Analysis Working Group, 2008.

[13] Kate Brannen, "Combat Brigades in Iraq Under Different Name," *Army Times*, August 19, 2010.

[14] Defense Science Board, *Fulfillment of Urgent Operational Needs*, Washington, D.C.: Office of the Under Secretary of Defense for Acquisition, Technology, and Logistics, July 2009.

Figure 6.2
Commanders' Decisions in Force Design

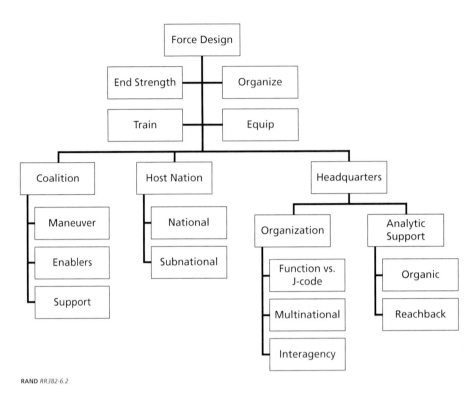

up to get what they need. Some units hedge, saving money until they get in theater to discover what they need."[15]

The way commanders choose to employ forces also drives changes in force design by the institutional force providers. One Infantry Brigade Combat Team (IBCT) deputy commander noted that his area of responsibility encompassed 6,000 square kilometers, and 12,000 Afghans.[16] This sort of operational pressure on unit designs led the Chief of Staff of the Army to direct a tiger team review of selected tables of equipment (Army requirement documents for deployable units) to ensure that they accurately reflected mission requirements for full spectrum operations. One of the "strategic level" issues the study identified was the need to

> define/clarify the doctrinal roles, missions, and expected doctrinal battle space area in unit designs for each BCT type (H/S/IBCT) . . . Consider a doctrinal template for a BCT's battle space, which would drive planning, factors for communications coverage, logistic functions, and land ownership by unit type. Currently,

---

[15]   Interview with former theater staff director.

[16]   Interview with former IBCT commander.

units are operating in a much larger Area of Operation than they were designed to control/influence.[17]

There are some ongoing analytic efforts to support service-level force design for IW (e.g., Support for Strategic Analysis, Total Army Analysis), but comparable efforts to meet the needs of theater commanders still suffer from important gaps.[18] Some existing models are largely agnostic about the composition of units deployed, instead assessing the number of personnel needed by geographic region based on levels of violence.[19] This may create a bias for selecting units that can maximize the suppression of violence in the short term rather than units designed to focus on longer-term security force assistance efforts or population-centric reconstruction efforts. This may be appropriate for some campaigns and not others, or in some areas or at various times in a specific campaign but not others, so this approach is not clearly generalizable across all IW/COIN or IW missions.

Force Employment

Force employment covers a very broad range of activities. Arguably, even the Army's six broad warfighting functions may not address all possible employment decision points in IW.[20] This section addresses a selected set of decisions based on input from commanders we interviewed. We divide these issues into kinetic and nonkinetic activities. In campaign design, commanders make decisions about how to best allocate limited resources across kinetic and nonkinetic activities. In this section we discuss how commanders seek to most effectively execute these kinetic and nonkinetic activities. Commanders were typically interested in analysis that would help optimize the use of coalition forces, or to predict enemy activities.

Choosing where to allocate units across the battlespace (e.g., KTDs) is a crosscutting issue that affects all kinetic and nonkinetic activities. Consideration was given to where the enemy was operating, population center locations, and higher-echelon campaign plans. Commanders at the BCT level would pay careful attention to the identification of optimal sites for new combat outposts, the reallocation of forces between outposts, and the consideration to close outposts when appropriate.[21] Figure 6.3 identifies a range of force employment issues that commanders believed to be important.

---

[17] Headquarters, U.S. Army, 2011b.

[18] DoD Directive, *Support for Strategic Analysis (SSA)*, Number 8260.05, July 7, 2011; U.S. Army War College, *How The Army Runs: A Senior Leader Reference Handbook, 2011–2012*, Carlisle, Penn., 2012.

[19] Interview with senior analyst.

[20] The six Army warfighting functions are mission command, intelligence, movement and maneuver, fires, protection, and sustainment. Headquarters, U.S. Army, *Army Operating Concept 2016–2028*, TRADOC Pam 525-3-1, Washington, D.C., August 19, 2010a.

[21] Interview with commander.

Figure 6.3
Force Employment

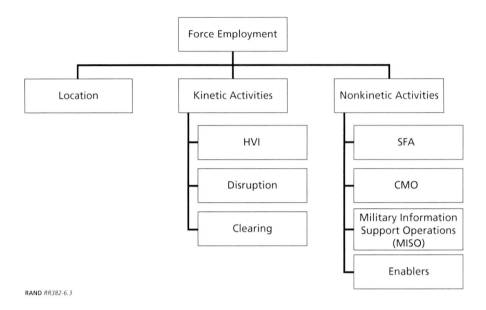

Kinetic Activities

Commanders we spoke with were interested in any capability that might help them predict enemy activities. They expressed particular interest in the ability to predict likely enemy ambush sites, infiltration and exfiltration routes, and logistical networks (e.g., finance, caches, routes). Commanders hoped to use this sort of analysis to inform how they employed their own forces.[22]

Kinetic activities ranged from narrow to broadly targeted. A narrowly targeted action might seek the detention of a single high-value individual, or interdict a single IED emplacer or emplacement team. More broadly scoped kinetic activities might include larger operations designed to clear an area of insurgent presence in order to enable sustained stabilization efforts. An intermediately scoped kinetic action might seek to disrupt enemy logistical routes or safe areas, without any broader intent to stabilize the area in the short term. Over the course of a deployment, a BCT might undertake only a handful of deliberate operations. Most of these were being initiated at the battalion level and below.[23] Commanders were also interested in understanding potential second-order and subsequent effects of kinetic actions. If a commander were going to conduct a heliborne assault, it would be useful to know what the civilian response might be.[24]

---

[22] Interview with commander.

[23] Interview with commander.

[24] Interview with commander.

Nonkinetic Activities

Commanders felt there was a "fault line" between the coalition's ability to defeat insurgents in the military and nonmilitary spheres, and believed that nonkinetic challenges were particularly difficult to understand. Nonkinetic activities largely fall into four categories: SFA, CMO, MISO, and use of enablers.[25]

A theater commander felt that MISO was the "least successful line of effort," noting that it was "difficult to win a war of ideas in a culture you don't understand very well."[26] IO targeted both elite and mass audiences. Elite audiences might be engaged through KLEs as commanders sought local support to enter an area. Support might be expressed through high accession rates into the host-nation security forces, or more passively through a lack of cooperation with insurgents. According to the commanders interviewed, a successful KLE requires identifying the correct key leader to engage, what their range of influence is, and what motivates them.[27] KLEs are also used to gain greater understanding of the operating environment, informing the conduct of both kinetic and other nonkinetic activities.

MISO activities might target mass audiences through a variety of media or kinetic or nonkinetic actions. IO activities might also be either deliberate or reactive. A deliberate IO campaign might include messaging the population about the history of Taliban atrocities. A reactive IO action might seek to explain the circumstances surrounding a civilian casualty. Commands must decide how to shape the appropriate message to the appropriate population segment through the right medium.[28]

CMOs are activities "that establish, maintain, influence, or exploit relations between military forces, governmental and nongovernmental civilian organizations and authorities, and the civilian populace."[29] CMO projects typically include school and well construction but could be as ambitious as facilitating mineral exports, land reform, or an election.[30] These activities might be undertaken to gain access to community, create jobs to temporarily reduce unemployment, or create an occasion to foster governance capacity.

Commanders were typically faced with two types of dilemmas when considering conducting a project: inclusion and capacity-building.[31] Commanders had to consider

---

[25] Most interviewees spoke of "IO" or "strategic communications" rather than MISO, a relatively recent doctrinal term, or psychological operations—which bears the unfortunate connotation of deception. For a discussion of doctrinal turbulence and MISO efforts in Afghanistan, see Munoz, 2012.

[26] Interview with commander.

[27] Interview with commander.

[28] Interview with commander.

[29] U.S. Joint Chiefs of Staff, 2011.

[30] Interview with commanders.

[31] For broader discussion see Stephen Watts, "Political Dilemmas of Stabilization and Reconstruction," in Paul K. Davis, ed., *Dilemmas of Intervention: Social Science for Stabilization and Reconstruction*, Santa Monica, Calif.:

how their projects might affect the interests of involved stakeholders. For example, they would want to know who benefited from the project and who was excluded. One special operations forces commander noted that the $5,000 he spent on a well for a mosque when he first arrived in an area—an effort that benefited only a few people associated with the mosque—would have had a more stabilizing effect if spent on projects supporting multiple stakeholders. The second dilemma was a tension between providing immediate services to a community that might result in immediate gains but undermine longer-term objectives. This was a genuine dilemma in that if the short-term objectives weren't met, the commander may not have the opportunity to seek longer-term solutions. Nearly every commander interviewed expressed great interest in acquiring analytic support to help determine how to best spend economic development funds, or whether to spend them at all.[32]

Many of the commanders and staff interviewed, particularly from the special operations community, expressed interest in understanding the root causes of IW/COIN conflict. Understanding why locals were participating in the insurgency, or failing to support the government, had implications for how commanders would choose to employ their resources. In many cases, commanders believed that local histories played a more important role in conflict than sectarian or ethnic ideologies, echoing David Kilcullen's theory of locally motivated insurgency proposed in *Accidental Guerilla*.[33]

In Afghanistan, one unit commander reported that his unit used a district stability framework to identify the agricultural sector as an opportunity area to engage the community. The command undertook an initiative to induce local farmers to start an agricultural cooperative, by making it a prerequisite of getting access to loans for farming equipment and seed. An agricultural cooperative is one where farmers pool resources to obtain efficiencies from economies of scale in certain activities. Loan applicants then had to explain their distribution and sales plan, which facilitated the unit's ability to map the community's human terrain. By having the Government of the Islamic Republic of Afghanistan's Directorate of Agriculture, Irrigation, and Livestock validate loan requests, the unit created a basis for the government to begin engaging the population in a way it previously had not been able to in this region. Once the cooperative was established, classes were held to teach farmers new agricultural techniques, supplemented by broadcasts that farmers could listen to on the radio. The analyst who developed the concept was inspired by the role of the National Grange Society in 19th-century America. The longer-term vision for the initiative was to create

---

RAND Corporation, MG-1119-OSD, 2012.

[32] Interviews with various staff. Some staff felt they already had adequate capacity to determine how to execute CMO projects, and were skeptical that additional support would bring much added value.

[33] Kilcullen, 2009

a constituency (i.e., local farmers) that was mobilized in support of the access to markets and capital that the government's presence could bring the community.[34]

SFA is the "unified action to generate, employ, and sustain local, host-nation, or regional security forces in support of a legitimate authority."[35] These activities in Afghanistan ranged from centralized at the theater level, through semicentralized at the division level, to radically decentralized training like Special Operations Forces training of Afghan Local Police. They also ranged from a focus on military capability to law enforcement, and from intelligence capacity to development of engineering capabilities.

Commanders at both the strategic and tactical levels felt that building host-nation security forces was a critical activity. Commanders needed to help organize, train, and equip host-nation forces, including recruitment and mentorship activities. At the national level, SFA included development of Ministry of Defense capacity. At the operational level, an operational commander noted that building interim security critical infrastructure (ISCI) local defense forces was the "smartest thing we did."[36] SFA approaches were seen as most successful when rooted in an understanding of local cultural norms. ISCI was thought to be a critical tool for harnessing the support of Afghans in Helmand province because it allowed them to protect their own neighborhoods and families in a way that was consonant with Pashtun norms, whereas the national Afghan National Police model was considered insufficient to meet Helmand's needs.[37]

Given the limited availability of a wide variety of enablers, commanders gave careful thought to their use. They paid particular attention to ISR and route clearance support. Commanders have an almost unlimited appetite for ISR, but access and processing capacity are always limited.[38] Commanders interviewed saw ISR as including a range of capabilities ranging from UAS to SIGINT, and they largely use these capabilities to identify and track enemy activities, though some expressed a desire to extend the application of these capabilities to the population. Typically, they would use these capabilities to help secure routes (e.g., identify IEDs being emplaced), support deliberate offensive operations (e.g., targeting Named Areas of Interest), or analyze the enemy network.[39] Though the employment of collection assets was considered important, commanders expressed greater interest in obtaining more collection assets rather

---

[34] Interviews.

[35] Headquarters, U.S. Army, *Security Force Assistance*, Field Manual 3-07.1, Washington D.C., May 2009b. This definition is evolving and may differ in various joint and service publications, but this version is sufficient for our purposes.

[36] Interview with commander.

[37] Interview with commander.

[38] Interviews.

[39] Interviews.

than obtaining additional analytic support that might optimize how they use the assets they already have.[40]

Maintaining mobility in both OEF and OIF has been a significant challenge, in some cases because of terrain, but largely because of the IED threat. Though vertical lift assets allow commanders to bypass these obstacles, they are another low-density/high-demand capability. Many commanders depended on the use of engineer units' route clearance packages (RCPs).[41] These are task-organized units typically consisting of engineers, armored vehicles, and other specialists; sometimes they consist primarily of general-purpose forces. Use of RCPs was complicated by the fact that their employment for one mission might render the capability inoperable for other missions. Commanders had to consider not only which units RCPs should be allocated to, but which missions they should be used to support. For instance, if RCPs are used for every KLE that comes up, they may not be available for deliberate clearing operations. Commanders were also concerned about the impact of route clearance activities on the population. In many cases, the population might appreciate the removal of IEDs, but when an RCP effort essentially shuts down a market road for several hours without discovering any IEDs, it's unclear how seriously the population takes the imposition. "Traffic got backed up behind us 80 vehicles deep (donkeys etc.). We held up traffic six hours, and really annoyed the locals. We didn't find any IEDs, but did we help those Afghans or not?"[42]

Vignette: Optimizing Force Size for IW

Interviewees tended to focus on strategic force structuring for IW, so the vignettes on this section are weighted accordingly. In this specific case, a joint military staff built and executed a wargame to identify the best possible force size for a large-scale IW mission. While this wargame was not created to support either OIF or OEF directly, it was relevant to both operations.

Analysis

This computer-aided simulation was intended to model requirements for all types of forces from all four armed services with the purpose of validating or informing preoperational planning expectations. Staff officers led the process and fed the information into the system, and they were also primarily responsible for divining the results of the simulation. While this was a "man in the loop" simulation, it was not an "analyst

---

[40] Interviews.

[41] Interview with commander. Also see U.S. Army Combat Studies Institute, *Wanat: Combat Action in Afghanistan, 2008*, Fort Leavenworth, Kan.: Combat Studies Institute Press, U.S. Army Combined Arms Center, 2008.

[42] Interview with commander.

in the loop" simulation. The absence of analysts affected the quality of the simulation and the value of its output.[43]

Data

Officers from each service began by providing their expected service requirements, determining all of the "above the line" forces prior to the execution of the simulation. In other words, the naval representative determined which capitol ships would be needed, and the ground representative determined how many divisions would be needed. The simulation was only intended to help determine what were termed *below the line* forces, like civil affairs units and special operations forces. The service officers were considered SMEs for the simulation; however, they also represented service equities and may have intentionally or unintentionally biased the simulation.[44]

Results

During the execution of the simulation, it became clear that the original model and structure underlying the process were unwieldy and perhaps unstable. To address this issue, the participants modified the underlying model during gameplay. This meant that the SMEs with specific equities were changing the rules of the game as it played out, undermining the objectivity and ultimately the value of the simulation. According to the interviewee, the simulation did not achieve its objectives and the output from the process was not deemed useful for planning.[45]

Issues

IW simulations tend to model some aspects of human decisionmaking, so they also tend to require input from SMEs during the preparation of the model, the execution of the simulation, or both. SME participation can add great value to a simulation by providing what some analysts termed a "reality check" to otherwise generic computer-aided decisions. Because all modeling begins with and rests on a set of assumptions about real-world conditions, SMEs can help provide the modelers with a better understanding of what those conditions are or might be. However, they also carry the baggage of subjectivity and preconditioned biases, they make factual errors, and sometimes they simply make poor decisions. Each of these problems can introduce error into the M&S, which in turn can lead to unrealistic deviations that may ripple throughout the entire process. In this specific case, the SMEs brought with them service biases that eroded the authenticity of the simulation.

Failure to include trained ORSAs or similar experts in the development of the model and the execution of the simulation also detracted from its overall effectiveness.

---

[43] Interview with senior analyst.

[44] Interview with senior analyst.

[45] Interview with senior analyst.

While OR and SA might or might not be generally applicable for all conceivable IW problems, an ORSA is probably the most qualified organic military asset for this kind of planning support effort. An ORSA could have refined the model prior to execution of the simulation, helped to preserve objectivity by adjudicating service equities, and defended the integrity of the model as the simulation began to unravel. While this vignette is anecdotal and draws on a single interview, it does indicate the value of ORSAs by offering an example of what might happen in their absence.

Vignette: Determine Campaign-Level Force Requirements

A commanding general tasked an in-theater IW staff to determine if the coalition could achieve victory with a smaller number of troops than were currently deployed. At the same time, the commander wanted to know how to best deploy the troops he had left. The small staff tasked with this order realized it could not handle the analysis alone, so it tapped into a reachback team to help develop a model and provide an answer. Some of the information in this vignette is generalized to preserve the anonymity of the interviewee.

Analysis

The combined forward-deployed and reachback analytic team took two different approaches to try to reach an answer: It looked at both force ratios and force density. The staff created a force-density model based on a selected set of historical cases, determining the ratio of friendly to enemy forces in each of these cases to identify a reasonable standard. Then, the staff used the Quinlivan model of 20 security personnel (e.g., friendly military, host-nation military, local police) per every 1,000 residents to form a basis for force-density analysis.[46] While the model itself incorporated other variables and SME inputs, these two calculations formed the basis of the analysis.[47]

Data

In order to determine force-ratio requirements, the team first had to develop a reasonable estimate of the number of insurgents in theater. The intelligence staff routinely provided an estimate of enemy strength, and the analytic staff capitalized on this esti-

---

[46] See James T. Quinlivan, "Force Requirements in Stability Operations" *Parameters*, Winter 1995. Quinlivan, a RAND researcher, developed this ratio based on an analysis of existing COIN cases to determine optimal force size for a prospective campaign, with the caveat that past is not necessarily prologue. According to FM 3-24, which measures forces used in the "area of operations," following the Quinlivan model would necessitate positioning 20 security personnel among each 1,000 residents, either as stabilizing forces or as forces to secure immediate perimeters to populated areas.

[47] Interview with senior analyst.

mate to provide its input into the model. Then, the team had to determine the size of the civilian population by region in order to execute the force-density calculation.

Data problems arose during this process that were endemic to all similar efforts in this and other IW theaters. Estimates of enemy forces may be considered reasonable enough to form the basis for analysis, but there is no way to provide a defensible estimate of accuracy to accompany these estimates in many—if not most—IW cases. In some estimative cases, the error range was 50 percent or more, while in others, estimates were provided down to single-digit precision without sufficient caveat.[48] Estimates across each historical IW case differ wildly in accuracy. For example, while there is a great deal of information available about the Viet Cong leading one to believe that the rough estimates of their forces might have been defensible, there is and was far less information about the Shining Path, Tupamaros, and other guerrilla forces.[49] Estimates of civilian populations were also problematic since the places in which the United States tends to conduct IW missions also tend to be places without adequate bureaucracy or transparency to develop valid, longitudinal census data over time. Even highly intensive, technical efforts to collect information on local populations like the Vietnam War era Hamlet Evaluation System are fraught with inaccuracy and misleading precision.[50] Technical approaches like LandScan have been used in both OIF and OEF to estimate population sizes, but these tools still provide rough estimates rather than replicable random sample census counts.[51]

Results

In this case, the commanding general delivered the problem along with part of the answer: The number of troops was a fixed independent variable. This meant that the analysts were able to focus on helping the commander understand how this force structure could be most effectively and efficiently deployed to meet his objectives. While they could not provide precise or accurate predictions, they were able to provide an answer that, at the very least, helped the commander think through his options. Even if he did not have blind faith in the model underlying the analytic output, it was useful as a starting point for his decisionmaking process.[52] There is always a danger in provid-

---

[48] These observations are made based on in-theater research in two separate IW theaters, and also examination of official intelligence estimates. Some of these estimates are publicly available while others remain unavailable to the public.

[49] Even with copious information, estimates of enemy strength were hotly contested during the Vietnam War.

[50] See Connable, 2012, Chapter 6.

[51] LandScan is a geospatial tool that uses spatial data and imagery analysis to develop ambient population estimates. Oak Ridge National Laboratories, LandScan web page, undated.

[52] Interview with senior analyst. In this case we do not have any other evidence to show that this process was effective or ineffective. However, this interview is supplemented by some direct observation by one of the RAND researchers and authors of this report.

ing this kind of forecasting: Structured analysis based on a nonvalidated process and incomplete and inaccurate data might mislead a commander. But there is no way of knowing how, specifically, the input from this particular analysis actually shaped the commander's subsequent decisions.

Issues

Two serious problems arose during this modeling and analysis process that are relevant for all force-structuring efforts. First, there is no verifiable standard ratio for either IW troop-ratio calculations or IW troop-density calculations. All forecasting and prediction is derived from past experience, or historical cases, so anyone undertaking these efforts must take into account that past is not necessarily prologue; it is a logical fallacy to extrapolate lessons from historical events for futures prediction. Instead, past cases can inform more detailed analyses of prospective or ongoing cases. The processes of selecting IW case studies to develop a ratio, of identifying representative ratios in past IW cases, and of making comparative assumptions regarding the relative value of various IW forces are all highly subjective. The range of disagreement in the literature on troop ratios and troop-density calculations, and also the heavy caveats associated with this literature, should leave anyone attempting to use these ratios for a real-world problem with serious reservations.

Even if a generally agreed-upon and verifiable ratio did exist for these kinds of calculations, the data required to populate an effective model are most often derived from poorly educated guesses, or questionable proxy standards, rather than direct observation. Census data from both Iraq and Afghanistan were more than a decade old by the time the figures were used in force-density calculations for OIF and OEF, and even then these data were highly questionable. In general it is unwise to extrapolate from old census data since war has the effect of displacing and disrupting normal life cycles and growth patterns. Even those census data acquired during a complex and prolonged conflict should only be viewed as incomplete and temporary snapshots rather than a basis for stable longitudinal analysis.

Yet despite these serious concerns, the analytic staff in this case was left with no choice but to produce a quantifiable response. The interviewee stated:

> Operations analysis in COIN is messy, slimy, OR work. The data are incomplete and contradictory, but you have to hang your hat on something. For example, the Steven Goode work on force density is one of the few pieces out there on the subject. He stated that we shouldn't use his calculations to model force density, but it is all that is available.[53]

---

[53] Interview with senior analyst. Steven Goode, "A Historical Basis for Force Requirements in Counterinsurgency," *Parameters*, Strategic Studies Institute, Winter, 2009.

In other words, if one is tasked with modeling force structure for a theater commander in an ongoing IW campaign, one must accept and make every effort to account for the serious flaws in force-structuring methods and data and then press forward. In this case, the analysts compensated for these problems by providing wide estimative ranges rather than precise figures. At no time were any of the analysts involved in the project under the illusion that their model or their analysis offered a pathway to accurate prediction. Thankfully, according to the interviewee, the commanding general was an educated consumer and able to appreciate the limitations of the approach and of the team's results. The interviewee also noted that reachback support for this project was not only invaluable, it was absolutely necessary.[54]

Vignette: Estimating Force Transition Requirements

A senior joint military commander asked his staff to provide a troop drawdown analysis for two separate IW campaigns. Specifically, he wanted to know the optimal size of host-nation security forces for both campaigns, and then how to most effectively draw down forces over time. Analysts on the staff created a model and then executed a wargame to answer this commander's question.

Analysis

The staff used the Quinlivan troop-density ratio to form the basis of their model. They added some cases to the existing Quinlivan study, then incorporated a force structure calculation based on 20 security forces for every 1,000 civilians for each theater. This calculation became the "engine for the wargame."[55] They built outward from this ratio, incorporating variables based on a review of IW literature and doctrine, SME advice, and previous efforts to develop similar models. The "man in the loop" wargame involved not only analysts, but also players who were brought back from deployment to participate and to inject realism into the inputs and outputs of the game. The staff also tried to model enemy activity, but since they did not know enemy strength, they used enemy violence as a proxy variable.[56]

Data

Multiple data sources were used to feed the model, but it primarily relied on the input of SMEs, the use of SIGACTs as a proxy for enemy presence, and estimates of the population. It was generally recognized that all data were incomplete and inaccurate to

---

[54] Interview with senior analyst.

[55] Interview with analyst.

[56] Interview with analyst.

an unknown degree, and that many of the variables used were proxy variables that also consisted of incomplete and somewhat inaccurate data.

Results

The model and wargame produced outcomes that were internally consistent and verifiable using the methods and data employed. However, this verification was only consistent within the wargame. The model began to "fall apart" as coalition troop levels were lowered during the withdrawal phase of the game. There was simply insufficient information at lower troop levels to sustain the game.[57] According to the analyst interviewed, IW becomes "a different kind of war" at lower troop concentrations (less than the typical 100,000-plus range for OIF and OEF).[58]

Issues

Nearly all of the force-structuring models developed to support OIF and OEF incorporated a range of variables and subordinate models in an effort to mirror some of the real-world complexity faced by IW decisionmakers. This approach is reasonable and perhaps necessary, although at some point the complexity of the model can cause it to collapse during computer-simulated gaming. Simpler pattern identification models might hold greater promise. For example, the CAA Force Level Analysis Counterinsurgency model revolved around an analysis of historical cases and identified a simple threshold ratio as a force optimization point: Hit the "right" number of combined troops, and violence starts to go down. However, even models that attempt to cut through complexity to find a simple universal truth are vulnerable to logical fallacies, poor data, and idiosyncrasies in historical cases.

Summing Up

As with campaign assessment, the strategic aspects of force structuring provide analysts with little in the way of verifiable methodology or data. Based on our literature review and interviews, there were no clear-cut analytic successes in support of predictive strategic force-structuring decisionmaking. Because it is predictive, this kind of analysis is inherently difficult for the analysts and dangerous for commanders. While the analysts in the vignettes listed here were careful to present their findings as broad estimates with noted caveats, there is always the danger that a commander will latch on to a quantitative prediction or simply demand a more precise answer to feed a key decision. Because strategic analyses are so complex and poorly bounded, commanders

---

[57] Interview with analyst.

[58] Interview with analyst.

share perhaps as much responsibility for the successful and reasonable outcome of strategic analyses as their analysts.

This does not mean modelers and analysts have made no progress toward exploring more effective models and analytic methods. Based on our literature review and interviews, CAA has taken the lead in force modeling and analytic efforts by developing multiple force-employment models and databases.[59] The development of multiple models and approaches is preferable to the existing doctrinal fixation on the Quinlivan force-density calculation, an approach that James Quinlivan himself does not recommend as a singular or predictive method.[60]

Analysts should also consider whether they should expect even reasonable accuracy from an effort to model such a complex and idiosyncratic problem like force structuring, at least at the strategic level. Instead, it might be worth investing more resources in operational and tactical analytic support to military staffs in their efforts to determine force structure requirements for IW. These more granular and specific efforts might be more accurate, they could be more reasonably precise, and they might even reach a point at which they could provide effective real-world (as opposed to simulated) support to decisionmaking. Some analysts have made these kinds of contributions in theater, but in an ad hoc manner. As one senior ORSA interviewed for this report noted, "Nothing will replace really good staff work in combat."[61]

---

[59] Smith, 2010; as well as multiple interviews with several CAA analysts.

[60] This observation is derived from a reading of James Quinlivan's collective works on force requirement analysis, and also from discussions with James Quinlivan.

[61] Interview with senior analyst.

CHAPTER SEVEN

Conclusions, Findings, and Recommendations

There cannot be a complete and accurate accounting of the contributions that modelers, simulations experts, and operations analysts have made to commanders' decisionmaking in OIF and OEF. Analysts' work is done quietly, in small offices, cubicles, and on field desks, and then typically delivered as only one of many contributions to a commander's overall decision process. Most service members, and even many commanders, are probably unaware of this work or even of the people making these contributions, and very few analyses are ever distributed in publicly available documents. The best evidence of successful analytic support is often anecdotal and idiosyncratic. Yet it is clear that modeling, simulation, operations, and systems analyses have helped commanders think critically about the COIN and IW environments, greatly reduced costs to the U.S. taxpayers, made logistics operations far more efficient, and in more cases than we can know, saved lives.

Was the Support Effective?

Some commanders interviewed for this report were enthusiastic as they described the contributions of modeling, simulation, and analysis to their efforts. They were particularly grateful for the support provided to their subordinate or tactical units.[1] Analysts were equally proud of these efforts, apparently with some justification. It was possible for us to identify some anecdotal evidence that specific logistics and force protection models, simulations, and analyses achieved analysts' objectives. There were few complaints from commanders or analysts that DoD was ineffective in providing analytic support to either tactical decisionmaking or to logistics or force-protection decisions. Instead, they tended to demand more resources for these efforts. One analyst interviewed for this report stated, "We don't make the decisions. We make the decisions better."[2]

---

[1] Most of the commanders and former commanders we interviewed had commanded at the battalion level or above, and therefore commanded subordinate tactical elements.

[2] Interview with junior analyst.

Interviewees expressed far less enthusiasm for the results achieved in support of campaign assessment and strategic force structuring. While the results of 115 interviews cannot be extrapolated to provide a generalizable reflection of all commander and analyst opinion across DoD, there appeared to be a consensus within this pool of interviewees that something was missing from the support to key strategic decisions. These opinions were also reflected in a preponderance of the literature sampled in Chapter Two. While we could find ample documentation of clearly defined, quantifiable, and effective support for analytic challenges such as transportation optimization and C-IED, the literature on campaign assessment and force structuring was sparse, conflicting, and generally inconclusive. Authors of various reports identified problems with strategic modeling and analysis and suggested theoretical paths to improvement, but they were not able to successfully bound these complex strategic problems in a way that was generally accepted across the community of experts. The most confident and clear-cut proposals either have not worked in practice (or have not shown clear evidence of success) or have only been applied tentatively in the months leading up to publication of this report. A possible reason for this in some cases might be that solutions to complex IW problems take a considerable amount of time to have an effect. Modelers and operations analysts working to address these strategic problems face a range of hurdles that they and others do not face when addressing tactical and more readily quantifiable problems like transportation optimization and C-IED analysis.

The lack of a consistent and generalizable theory of COIN and IW strategy; the sheer number of possible interrelated variables that must be accounted for at the strategic level; and the paucity of available, complete, and accurate data all undermine the accuracy and effectiveness of processes that depend on a clearly framed problem; a bounded set of variables; and plentiful, accurate, quantitative data.[3] And while some criticism of strategic analyses is clearly warranted, the ability to show that these analyses had a positive causal impact on decisionmaking and outcomes is exponentially more difficult than for simpler, tactical analyses. In other words, finding metrics to show the *value* of strategic analyses is as difficult as finding metrics to *support* strategic analyses.

Continuing Analytic Challenges

Returning to the ten-step ORSA process described in Chapter One (and reproduced in Figure 7.1), it is possible to see where these analytic challenges undermine the scientific approach to supporting commanders' key strategic decisions. There are limits to

---

[3] In many ways, this lack of data is understandable. Commanders collect data to support operations and not analysis. In an ideal world, the analyst would determine what data are needed to support a study, designate a control group and a treatment group and apply sound statistical analytical methods to arrive at a solution. In IW, this is impossible.

Figure 7.1
The Operations Research/Systems Analysis Scientific Approach to Problem Solving

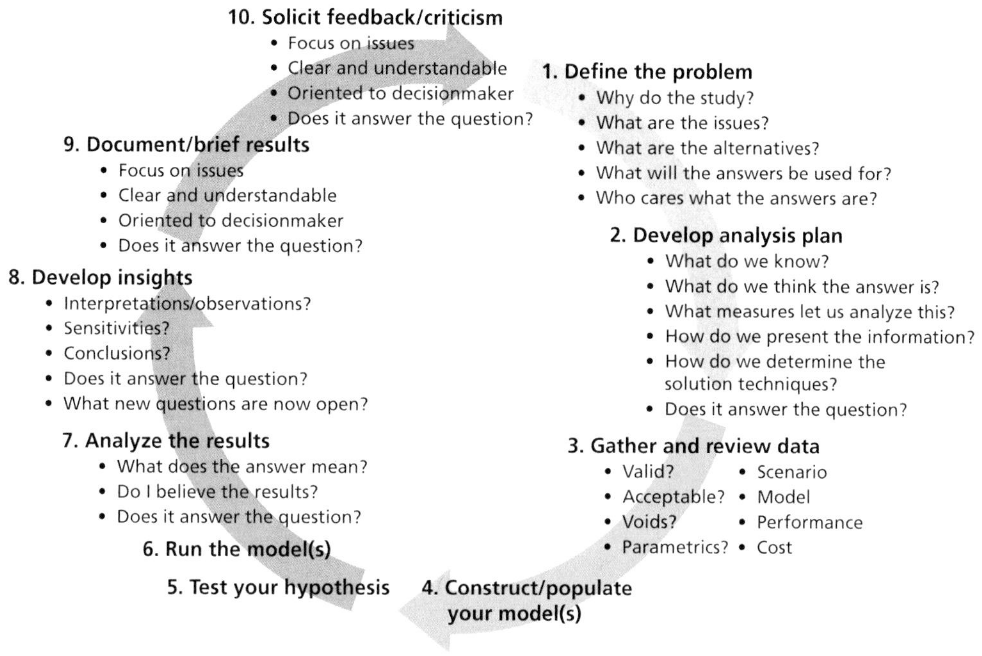

10. Solicit feedback/criticism
- Focus on issues
- Clear and understandable
- Oriented to decisionmaker
- Does it answer the question?

9. Document/brief results
- Focus on issues
- Clear and understandable
- Oriented to decisionmaker
- Does it answer the question?

8. Develop insights
- Interpretations/observations?
- Sensitivities?
- Conclusions?
- Does it answer the question?
- What new questions are now open?

7. Analyze the results
- What does the answer mean?
- Do I believe the results?
- Does it answer the question?

6. Run the model(s)

5. Test your hypothesis

1. Define the problem
- Why do the study?
- What are the issues?
- What are the alternatives?
- What will the answers be used for?
- Who cares what the answers are?

2. Develop analysis plan
- What do we know?
- What do we think the answer is?
- What measures let us analyze this?
- How do we present the information?
- How do we determine the solution techniques?
- Does it answer the question?

3. Gather and review data
- Valid?   • Scenario
- Acceptable?   • Model
- Voids?   • Performance
- Parametrics?   • Cost

4. Construct/populate your model(s)

SOURCE: The Army Logistics University, 2011, p. 5.
RAND *RR382-7.1*

the applicability of any analytic approach. While this cyclic model is intended to be a flexible process that individual analysts should modify to address each problem, it ultimately depends on analytic clarity and good data. The contradiction between the theory of this approach—one that can also be generalized to encompass many non-ORSA modeling and simulation efforts—and the realities of COIN and IW at the strategic level were made obvious in the literature and in our interviews. Yet while this problem has been acknowledged in various quarters, DoD has neither clearly addressed this inherent dichotomy nor provided modelers and analysts with a means to resolve it.

By comparing the COIN and IW problem sets, available data, the ill-structured nature of strategic objectives, and the difficulty in identifying key decisions with the ten-step process, it becomes clear that the process has limited applicability for strategic campaign-assessment, and force-employment decisions. Few of these steps can be addressed for either problem, and of those that might be, fewer can be addressed with the kind of clarity and consistency necessary to support a scientific approach to analysis. The analytic community might benefit from the development of new approaches to these kinds of ill-structured problems.

Commanders and analysts have identified a range of gaps in modeling and analytic support to key decisions in COIN and IW. While the lists of gaps provided here might not be complete, they offer at least a starting point from which the analytic community and DoD can begin to move forward. Simply plugging these gaps with more data collection or more analyses, however, may deepen preexisting fault lines between the commanders' needs and the analysts' ability to address those needs. The following sections provide specific, itemized findings derived from our research, and recommendations to address those findings that offer opportunities for improvement.

Findings

While our research addressed both COIN and IW, our findings explicitly reflect COIN experience in OEF and OIF. While these two campaigns reflect a subset of IW, we note that IW encompasses a broader spectrum of operations. Further, DoD may execute COIN in ways not reflected in OEF and OIF. For example, U.S. adviser support to the Salvadoran Army in the 1980s—limited to 55 advisors at any one time—could technically be described as a COIN mission.[4] While current U.S. doctrine (FM 3-24 and Joint Publication 3-24)[5] tends to describe COIN as an extensive whole-of-government campaign involving hundreds of thousands of military and civilian participants, readers of this report should note that COIN and IW can be executed at varying scales and across a range of environments that will present unique challenges. In the wake of OEF and OIF, such large-scale campaigns will probably be less likely and smaller-scale efforts will be more common. Therefore, some of these findings are generalizable across COIN and IW, while others may not be applicable across the full spectrum of IW decision support.

- **Tactical, logistics, and force-protection support has often been effective.** There is clear, empirical evidence of the success of modeling, simulation, and analysis in support of commanders' tactical, logistics, and force-protection decisions in OIF and OEF.[6] Analysts in DoD and in the nongovernmental analytic community have demonstrated ingenuity and innovation, generating new approaches, methods, models, and tools.
- **Conversely, there is little evidence that strategic, campaign assessment, and force-structuring analyses have been successful.** The success of modeling, simulation, and analysis in support of strategic, campaign assessment, or force-structuring decisions in OEF and OIF is less clear, and anecdotal evidence sug-

---

[4] See Paul P. Cale, "The United States Military Advisory Group in El Salvador, 1979–1992," thesis, U.S. Army Command and General Staff College, 1996, p. 13.

[5] Headquarters, U.S. Army, 2006d; U.S. Joint Chiefs of Staff, *Counterinsurgency Operations*, Joint Publication 3-24, Washington, D.C., October 5, 2009c.

[6] "Tactical" encompasses some tactical-level logistics and force-protection support. Some of the logistics and force-protection successes have been demonstrated at the operational and strategic levels of effort.

gests that efforts in these areas have left many commanders and staff officers dissatisfied. The complex COIN (and IW) environment makes it difficult to prove the effectiveness of even the best of these efforts. Strategic pattern and trend analyses can show some meaningful correlation, but rarely causality. Even effective correlation analyses (e.g., showing a decrease in violence during bad weather in Afghanistan) do not clearly support effective strategic decisionmaking.

- **It is not clear that OR and SA are applicable for all IW analytic problems.** Analysts have clearly had great success in applying the OR approach to quantifiable COIN and IW problems. They have been less clearly successful in applying the OR approach to nonlinear, complex COIN and IW problems that do not lend themselves to quantification. Based on our interviews with commanders and on previous RAND research on COIN and IW assessment, analysts appear to be generally competent and professional—this lesser effectiveness seems to reveal the inapplicability of the OR *approach* to some commanders' key decisions in IW rather than a failure of individual analysts. In some cases, other analysts (e.g., intelligence analysts, sociocultural analysts) have compensated for this gap, but DoD has not addressed this issue in depth.

- **Most COIN and IW decision support derives from simple analyses, not complex modeling.** While there have been extensive efforts across DoD and the supporting community to develop and employ complex models and simulations in support of OEF and OIF, most decision support comes in the form of rather simple analyses produced by forward deployed analysts. These analysts use basic automation tools such as Microsoft Excel and Access far more often than complex tools like those identified in the appendix. In some cases, analysts presented data in the form of time-series charts or percentages rather than as finished analytic products that included the analysts' professional, narrative interpretation of the data.

- **Reachback support for COIN and IW is useful, but its role is limited.** Many commanders and analysts praised the role of reachback support for OEF and OIF, particularly for complex and long-term decisions. However, most of those cited and interviewed also noted that this support is bounded by the relative lack of situational awareness in remote analytic shops, longer production time-lines, and the general inability of reachback analysts to support decisions within a 72-hour window.

- **Some commanders are insufficiently prepared to use analysts or analyses.** While some senior commanders are trained ORSAs—like General John Allen, who commanded in both Iraq and Afghanistan—many other commanders have a limited understanding of the capabilities, limitations, and demands of analysis. ORSAs we interviewed often stated that they had to explain their roles and capabilities to both commanders and staff. This hindered analysis in many cases, but it also gave some of the more entrepreneurial analysts the opportunity to do some excellent initiative-based work.

- **Commanders have trouble articulating their needs, but analysts tend to be self-motivated.** The complexity of COIN and IW makes it hard for commanders to clearly articulate their decision-support needs to analysts, and the dynamic nature of COIN and IW often imposes changes in these needs that undermine consistent analysis and assessment. However, most analysts were able to compensate for this lack of clear and consistent direction by determining requirements and generating analyses on their own initiative. This kind of self-initiated work is commendable and was often successful.

- **Many recurring, periodic reports had little value and sapped analytic capacity.** Unfortunately, the ability of analysts to take initiative was often constrained by the requirement to generate recurring reports, not all of which were clearly applicable for decision support. Once a report was generated and deemed valuable it was often demanded on a recurring basis regardless of whether it had recurring value. Recurring report generation is time-consuming and can overwhelm individual analysts and teams of analysts, reducing their ability to provide relevant, specifically targeted analytic support over time.

- **Data quality in COIN and IW are generally poor and inconsistent.** All literature sources and interviewees cited in this report noted the poor quality of data in IW. In OIF and OEF, data generally were incomplete, inaccurate, and inconsistent. This stems in great part from the fact that in nearly all cases, data are reported to support operations, not analysis. Worse, there was no clear way for analysts to determine the degree to which aggregated data sets were complete or accurate, particularly as aggregation was compounded. Data-quality issues were sometimes manageable at the tactical level, but rarely at the strategic level.

- **There is no clear understanding of what is meant by analysis or assessment in COIN and IW.** DoD provides a range of complex, overlapping, and sometimes contradictory definitions for these terms. Non-DoD and academic definitions contribute to this confusion. Inadequate definition undermines the ability of commanders to understand what support is available and which type of support might be best applied to specific problems. What is often described as analysis is actually assessment or simply data presentation. Both analysts and commanders demonstrated how this confusion undermined effectiveness.

- **Simulation, or wargaming, is useful for decision support in IW but has limits.** Simulation has helped analysts think about the challenges associated with complex and seemingly intransigent problems like operational force structuring, and it has helped prepare commanders and staffs for deployment to OIF and OEF. Tactical simulation has been useful for training and also in support of real-world problems like logistics and force protection. However, the complexity of the

IW environment and the lack of good, consistent data preclude the use of simulation as a real-world, real-time decision support tool at the strategic level.

Recommendations

This section offers recommendations to address those findings that offered opportunities for improvement or capitalization. It offers recommendations to three separate groups, but some of these recommendations may have applicability for more than one group.

For the Department of Defense

We recommend that DoD modify COIN and IW doctrine to include the provision that the execution of campaign and strategic assessment be included in the planning process. In COIN and IW, success in achieving objectives is difficult to assess because most often, operations are population-centric, and gauging how well the population responds to ongoing operations is extremely difficult. Whether the joint community retains the current metrics- and measures-based approach to assessment or adopts another approach, an approach should be officially integrated into planning and operations.

In addition, ISAF's 2012 approach to campaign assessment should be considered for inclusion in COIN and IW doctrine, and new doctrine should more clearly acknowledge the challenges of IW force structuring. Instead of providing a singular approach or set of methods for these complex challenges, doctrine should offer a range of approaches and methods from which commanders and analysts can choose to fit specific situations.

DoD should consider whether to continue drawing almost exclusively from the ORSA community to provide strategic decision support or if it can provide an alternative capability to operational commanders. One alternative to relying primarily on ORSA support would be to create multidisciplinary assessment teams. Several commands have taken this approach in both Iraq and Afghanistan, although ORSAs tend to constitute the majority of analysts outside of the intelligence staff. Matching ORSAs with analysts who might have different or less-structured approaches to problem-solving might generate mutually reinforcing improvements in key decision support. For example, combining an ORSA with a social scientist, an intelligence analyst, a civilian adviser, and a graduate of an advanced planning school (like the Army School of Advanced Military Studies or the Marine Corps School of Advanced Warfighting) could produce a synergistic analytic capability that would be more capable of addressing the fluctuating and often ill-defined needs for commanders in IW.[7]

For Modelers, Simulations Experts, and Operations Analysts

Modelers and analysts should consider identifying the limits of various approaches, methods, and tools. This will help commanders and other consumers understand where and how modeling, simulation, and analysis can be helpful, and where the complexity of a problem begins to impose limits on efficacy. Specifically, analytic and assessment reports should include clear qualifying remarks as to the completeness and accuracy of the data used. If analysts cannot determine these qualifications, reports should make this limitation abundantly clear. Analysts should continue to seek ways to incorporate qualitative data into analyses and assessments, and also to find new and innovative ways to situate their analyses within the holistic context of the overall campaign. The analytic community should continue to press DoD to incorporate those innovations that have proven so successful for tactical, logistics, and force-protection problems into technical manuals, doctrine, and organic capabilities.

For Commanders

Commanders who understood the capabilities and limitations of their analytic support staffs seemed to be more effective in using these assets than less-informed commanders. Therefore, it would behoove all commanders to familiarize themselves with the capabilities and limitations of their analysts and of the methods and tools they employ. This familiarization should encompass not only ORSA, but also all other available analytic assets.

Perhaps more important, commanders should make every effort to provide analysts with clear articulations of their key decisions. Analysts were more effective in supporting commanders when they had a good understanding of the commanders' needs. While this kind of clear articulation is often elusive in IW, any effort to communicate needs will be helpful.

Commanders should also encourage initiative-based analyses and assessments, and also periodically revisit the requirements for recurring reporting to ensure these reports are necessary and not placing an inordinate strain on their available analytic assets. Commanders should make use of reachback support, but should tailor their requirements and expectations in accordance with our findings.

Finally, commanders should avoid including the desired answers in questions posed to analysts and insist on objectivity in their assessments. Often, recommended

---

[7]   One example of this approach can be found in the ISAF Joint Command's Information Dominance Center. This center was intended to combine a range of intelligence specialists, ORSAs, planners, and other staff officers in an open-plan environment. Execution of the IDC concept was hampered by staffing problems and other issues associated with the development of a new working concept. However, the concept itself is worthy of consideration for future operations.

decisions resulting from an analytic effort will contain caveats and assumptions that commanders should not ignore. It is tempting to discard inconvenient caveats and embrace welcomed assumptions when the decision recommended is in line with the desired outcome, but doing so paints an incomplete picture of the analytic context in which the recommendation was made.

Review of Selected Models and Simulations

In this appendix, we take an in-depth look at a sample of models and simulations that have been used to support decisionmaking in OIF or OEF. In some cases, these applications were described as having successfully supported decisions. Most often, though, both commanders and analysts viewed complex models and simulations as resource- and time-intensive. For this reason, many models and simulations like those described below helped analysts and commanders think through the outcomes of their decisions, rather than use them for operational planning. Forward-deployed analysts were less likely to have the time or resources available to use complex M&S than analysts working from a reachback role. In general, teaming forward-deployed analysts with a robust reachback capability appeared to offer the best opportunities to take advantage of these types of models and simulations.

The Peace Support Operations Model

The PSOM is a "faction-to-faction, turn-stepped, cellular geography, semi-agent-based model" developed in 2006 by the Defence Science and Technology Laboratory (Dstl) [sic] of the United Kingdom Ministry of Defence (UK MoD).[1] It is a wargame that incorporates the broad array of civil and military components of Peace Support Operations (PSO), including crisis management and security and stabilization operations, to determine how these factors affect campaign outcomes. UK MoD generated the requirement for such a wargame based on the need to obtain a greater understanding of stabilization and COIN based on their involvement in Iraq and Afghanistan.[2] Dstl designed the PSOM such that it is consistent with both the U.S. and UK definitions of IW and PSO. By working closely with the U.S. Joint Staff J-8 Warfighting Analysis

---

[1]   Body and Marston, 2010.

[2]   Jeff Appleget, "PSOM Overview and Peacekeeping Operations Assessment Using PSOM," Briefing delivered to the Naval Postgraduate School, Monterey, Calif., October 2011.

Division (WAD), Dstl ensured both countries' essential IW, COIN, and peacekeeping doctrine was incorporated into the PSOM.[3]

The PSOM is designed for implementation at the strategic level to address policy and theater-specific questions regarding civilian perceptions of ongoing operations. Two mutually dependent levels of war are examined and operate simultaneously—the Strategic Interaction Process (SIP) and the Operational Game. The SIP simulates the strategic decisionmaking of political and military leaders, while the operational game translates these decisions into campaign effects.[4] Activities are carried out by formations (military battle groups) and teams (civilian reconstruction groups) in a "man in the loop" construct, with the goal being to demonstrate how policy decisions, strategic decisions, and operational effects are all connected.[5] Thus, the PSOM is intended to model a society's transition from political anarchy to self-sustainment and back again, if necessary. Campaign activity is modeled along the following LOOs: military, economic, social, and political.[6]

Background

The PSOM was designed to help decisionmakers consider the populations affected by IW, but it was not intended to be a predictive model. Rather, it is a tool to help commanders understand population as the center of gravity. It has been used as a predeployment or training tool at NATO and in countries including the United Kingdom, United States, Canada, and Germany to prepare staff for the challenges of IW. It has also been used to conduct real-world assessments for scenarios in Iraq in 2007 and Afghanistan in 2010, 2011.[7] The PSOM operates under the assumption that a population's behaviors, attitudes, and beliefs change slowly. Accordingly, time is measured in months and years instead of days while running the wargame. Figure A.1 shows the step-by-step process of a PSOM "turn," and how the PSOM attempts to measure both the mission's intent as well as any unintended second- and third-order effects on the population.[8]

Each game turn is one month and includes the following materials:

- a map of the area with specified grid sizes
- the factions playing: red, blue, green, brown, white
- starting conditions: contextual information and intelligence
- player stance: the player's plan as viewed after the last turn

---

[3] Body and Marston, 2010, p. 72.

[4] Body and Marston, 2010, p. 70.

[5] Body and Marston, 2010, p. 70.

[6] Appleget, 2011.

[7] Appleget, 2011.

[8] Appleget, 2011.

Figure A.1
The PSOM Sequence of Events

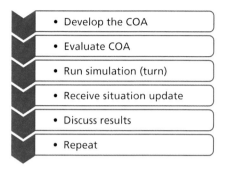

SOURCE: Adapted from Appleget, 2011.
RAND *RR382-A.1*

Inputs, Process, and Outputs

This is a "player in the loop" wargame. During the input process, players may interact with each other to simulate diplomatic and political activity in the real world. The midlevel analyst participating in the wargame provided insight into the data that fed into the wargame. For the 2011 exercise he participated in, this information came from planners in all regional commands in Afghanistan. In general, inputs to the model consist of simple representations of the following:[9]

- all protagonists in the campaign (factions)
- the environment
- the infrastructure
- the population
- the economy and employment
- political level interactions
- military units and combat
- reconstruction and civilian units
- human and 'soft' factors.

Once the input is set, the adjudication process begins. Plans are scripted, so units and others proceed according to the plan for the full month. According to a midlevel analyst with experience testing the PSOM in Afghanistan, outcomes of the interactions among the players are adjudicated stochastically. For example, if the plan calls for coalition forces to move to a location where the Taliban have also moved, the resulting number of casualties on both sides is adjudicated probabilistically.

---

[9] Jon Parkman and Nathan Hanley, *Peace Support Operations Model Functional Specifications (PSOM-FS)*, Dstl/TR28869/1.0a, August 13, 2008.

Outputs are measured in:

- **civilian casualties:** An algorithm calculates the effect on the civilian population of Red/Blue engagements.
- **security:** The local population's perception of how secure they feel. Security is functionally related to civilian casualties: As casualties increase, security decreases.
- **consent:** This is defined as the proclivity of the population to do what they are told.
- **initiators of kinetic events:** This assesses how ready the ANSF is to take charge. If most of the kinetic activity is initiated by coalition forces, then the ANSF is considered not ready.
- **readiness levels:** Readiness of ANSF is measured using manpower, leadership, and experience levels.

Analysis of the PSOM Process and Approach to Decision Support

According to the Dstl, PSOM has been used regularly during OEF to support planning decisions regarding force deployment. Two PSOM planning conferences were held in Afghanistan in 2011 to support U.S., UK, and NATO planning that included Dstl civilians, Afghan government personnel, and NATO military planners, among others. At these conferences, the PSOM was used to simulate planning, execution, and assessment of real-world operations. The goal was to provide senior military and civilian decisionmakers with clear direction as they shaped their force deployment strategies in Afghanistan. The ISAF Joint Command found value in these conferences and used the insights gained during one of the sessions to inform decisions regarding high-level campaign objectives, foreign troop commitments, and the transition of responsibility to Afghan security forces, according to Dstl, which also points out that the advantage of PSOM is to conduct this type of future forecasting without putting troops in danger.[10]

Several analysts interviewed for this report described the PSOM as useful overall, but most had only a cursory understanding of the wargame and its objectives. The PSOM is ambitious in its scope and attempts to model a difficult notion; namely, the perceptions of the population whom IW activities will affect. The authors of the simulation have knowingly taken a risk in attempting to model such a complex problem: Data dealing with popular support, such as opinion polling or surveys, are often flawed and rarely collected with uniformity across the selected population. It is not clear that one can draw accurate representative samples in Afghanistan, a country that lacks accurate census data and suffers from significant population disruption and divisive heterogeneity. None of the real-world data from an IW/COIN campaign should be considered either fully accurate or complete, so even if the model itself has perfectly

---

[10] Email interview with Dstl representative, December 5, 2012.

recreated the environment—a claim the authors of PSOM wisely do not make—the outcome cannot accurately represent the real world.

Consumers who understand these restrictions and accept that "a model is just a model" are most likely to find PSOM and similar wargames valuable. Again, the PSOM was not intended to provide real-world prediction. Commanders who take PSOM outcomes literally are likely to be misled by simulation results, or lose faith in the PSOM approach to modeling the complex IW environment. The midlevel analyst who was part of the 2011 PSOM run-through said generals would disregard the outcome of the PSOM when they knew the information it provided was incorrect.

Recommend Tool for Future IW Use?

The PSOM is a time- and data-intensive wargame that requires many different players to represent various scenarios. As a result, it is difficult to obtain results quickly. However, it is designed to help commanders think through the various ways a campaign will affect the native population, which makes it a useful planning tool. One analyst described it as a "useful conversation piece" to move operational scenarios along. As with any model, simulation, or analytic tool, the PSOM is best implemented for the purpose it is designed to support: planning for commanders entering a battlespace where civilian and military actions will often overlap, and considering the possible outcomes of these interactions with respect to the overall success of the campaign. Using it as a predictive tool or direct input to specific operational decisions will not yield the intended results. With additional validation and careful consideration of the data that feed the model, the PSOM and similar wargames are poised to provide generalizable insight into how operations can be tailored to minimize their negative impact on civilians and improve their quality of life.[11]

Athena

Athena combines the efforts of several predecessor models to simulate the interaction of civilian populations with armed forces, particularly during stability and reconstruction operations. The direct predecessor of Athena is the JNEM, developed by the National Aeronautics and Space Administration's Jet Propulsion Laboratory (JPL) in response to a 2002 requirement established by the Army's National Simulation Center. After JPL introduced its first version of JNEM in 2005, the model was used in several training federations, including Unified Endeavor 2006.[12] JNEM relied on external players and

---

[11] This assumes that improving quality of life and changing public opinion are the keys to success in IW/COIN, but this assumption is disputed. At the very least, any consumer of the PSOM or similar wargames would be wise to consider them as a single input rather than a holistic response to complex questions.

[12] Henry, 2009.

federations for its simulation inputs and typically involved the execution of only brief scenarios. By 2009, JPL delivered its first version of Athena, which sought to create a single-user, decision support tool that examined multiyear planning perspectives.[13]

As of January 2012, Athena had undergone three separate revisions. The current version at that time, Athena 3, modeled six separate but interrelated environments, or "areas:" ground, demographics, attitudes, politics, economic, and information. Each area consists of several models and algorithms that depend on inputs and outputs from each of the other areas. Some of these models include the Generalized Regional Attitudes Model, with inputs from the Driver Assessment model, which both have their roots in JNEM; the Mars Affinity Model; the Athena Attrition Model; an economic model of three sectors of Computable General Equilibrium using Gauss-Seigel algorithms; and the TRAC human intelligence methodology.[14] Figure A.2 depicts the relationships among the six modeled environments to form the overall Athena simulation environment.

Figure A.2
Relationship Between Modeling Areas in Athena

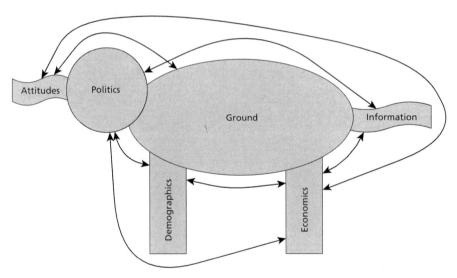

SOURCE: Duquette, 2012, p. 13.
RAND RR382-A.2

---

[13] For a more detailed history, see Robert G. Chamberlain and William H. Duquette, "Athena in 2011," paper presented at the Military Operations Research Society Special Meeting on Operations Research Methods in Support of Countering Transnational Threats in McLean, Va., December 12–14, 2011.

[14] See William H. Duquette, "Athena User's Guide, Athena A&RO Simulation, V3," JPL, January 2012.

Data

To populate the Athena model, a selected analyst must build each scenario from the ground level, following a 14-step checklist intended to help analysts replicate the modeled environments. The analyst is responsible for all of the inputs, so the data are "qualitative and inexact."[15] According to a midlevel operations research analyst with experience using Athena, it takes four months to run a complete cycle. The first month is spent gathering data (validated by SMEs from the Intelligence Community and U.S. combatant commands) and entering them into the model. Inputs include defining actors and their resources, beliefs systems, and strategies; civilian, force, and organization groups and their satisfaction and cooperation levels; the "neighborhood"; and environmental situations. After inputting these parameters, the operator allows the simulation to run and can interject with "magic" capabilities, adjusting attrition to account for noncombat deaths, attitude changes, and environmental situations when he or she sees fit.

Results

To date, two classified studies have been produced using Athena to the end user's satisfaction. However, the results of these studies are not publicly available. The midlevel analyst familiar with using Athena said customers have told the analysts they are happy. In his experience, the product disappears as it moves up the chain. If customers stopped asking for assistance, we would consider that negative feedback.[16]

Athena seeks to solve operational and strategic problems rather than tactical ones. For example, Athena could be applied to anticipate consequences of force activity, or the second- and third-order effects upon noncombatant groups and their potential responses. To discern these outcomes, Athena users compare multiple COAs simultaneously to discern potential outcomes from political, military, economic, and social interventions. This also helps answer questions regarding the local populations' conditions (e.g., their standards of living and unemployment rate) as well as when and how to withdraw U.S. troops.[17]

TRAC Irregular Warfare Tactical Wargame

TRAC designed the Irregular Warfare Tactical Wargame (IWTWG) in 2008 to fill the M&S gap in DoD's ability to inform senior leaders' IW decisions. It was specifically designed to advise the Army on equipment and personnel to support the Army

---

[15] Chamberlain and Duquette, 2011, p. 7.

[16] Interview with midlevel analyst.

[17] Interview with midlevel analyst.

Force Generation Process.[18] This game was an outgrowth of the PSOM, which is designed for the strategic and operational levels. TRAC developed the IWTWG with CNA, TRADOC G2, Naval Postgraduate School, Texas A&M University, Argonne National Labs, Charles River Analytics, the University of California at Davis, and the Army Material Systems Analysis Agency. It focuses heavily on human behavior and the social and cultural factors at play during IW conflicts.[19]

The IWTWG envelopes methods, models, and tools (MMT) into a "man in the loop" wargame to provide operational context to battalion and company commanders about an area of operations. This context includes information about the interactions between ground forces and civilians, as well as who the key leaders are in the region and what type of infrastructure is available. The IWTWG also provides information regarding civilian reactions to events on the ground for players to use in their COAs. It consists of three components:

- **operational wraparound:** an operational-level tabletop wargame conducted to provide context for the larger operation. The wraparound simulates command and control functions and incorporates elements of the local infrastructure and population atmospherics to inform the tactical wargame.[20]
- **planning, adjudication, and visualization environment:** a "man in the loop" tactical-level tool that allows players to plan, assess, and record tactical-level wargame inputs. In this component, players have access to key local leaders and population and infrastructure interaction information that will assist commanders in carrying out daily operations.
- **cultural geography and nexus network learner models:** designed to provide players with population responses to events carried out through the course of the game. Namely, they provide information regarding effects of infrastructure changes and tactical operations on the population, as well as the outcomes of key leader engagements.[21]

---

[18] The Army Force Generation Process was implemented in 2006 in response to the high demand for ready units and quick turnover between deployments. The Army defines this process as "the structured progression of unit readiness over time, resulting in recurring periods of availability of trained, ready, and cohesive units." Headquarters, U.S. Army, *2012 Army Posture Statement, Addendum G—Army Force Generation (ARFORGEN)*, Washington, D.C, 2012.

[19] TRADOC Analysis Center, "TRAC Irregular Warfare Tactical Wargame Update," briefing for the IEA 1448 Meeting, March 14, 2012.

[20] According to the TRADOC Analysis Center briefing for the IEA 1448 meeting in March 2012, these atmospherics are "the players' perception of the population response to [their actions] during interactions."

[21] List derived from TRADOC Analysis Center, 2012.

Data

According to a midlevel operations research analyst familiar with IWTWG, TRAC uses social science SMEs to validate the data that feeds the wargame. These data include information regarding the population, key leaders, local infrastructure, and perceived population response to Blue force actions.[22] The SMEs come from TRAC–Ft. Leavenworth, TRAC–White Sands, and the Naval Post Graduate School. They review cultural narratives that describe how people's attitudes have been changed by the behavior of various actors. The SMEs complete a survey review of the narrative from the perspective of the culture in question. TRAC uses triangulation to judge the accuracy of the narrative; if three SMEs agree, that is "as good as it gets." They also receive Gallup poll data from Afghanistan and have access to CAA's database of surveys from OEF.[23] Establishing the wargame for a new area of operations could take about nine months, including time to get all the data and triangulate them. TRAC spent three years prototyping the wargame. During the fiscal year 2011 Prototype II improvement phase, additional intelligence data were added to improve the information available to the players in the operational wraparound component.[24]

Analysis and Outputs

In October 2011, TRAC led the Prototype II Wargame, an OEF-based scenario with a brigade-level operational wraparound. Seven agencies participated in the game, including multiple Army intelligence units. The goal of the game was to assess how adding a Company Intelligence Support Team at the company level would assist the commander in influencing the local population. The game examined two infantry battalions with four maneuver companies each who were all conducting COIN operations, one with four support teams and one with no additional intelligence support.[25] The results of this game are unavailable for analysis. However, the intended output of the IWTWG is to determine how the civilian population will respond to Blue force actions and how infrastructure changes will affect the population, and to predict the results of interactions with key leaders.[26] The issues that applied to the PSOM, issues with data, and concerns with accurately modeling a complex environment also apply to Athena and the IWTWG.

---

[22] TRADOC Analysis Center, "TRAC Irregular Warfare Tactical Wargame Update," briefing for the IEA 1448 Meeting, March 14, 2012.

[23] All of the work discussed by the subject was regarding their 2011 Afghanistan scenario; interview with several mid-to-senior-level analysts.

[24] TRADOC Analysis Center, 2012.

[25] TRADOC Analysis Center, 2012.

[26] TRADOC Analysis Center, "TRAC Irregular Warfare Tactical Wargame Update," briefing for the IEA 1448 Meeting, March 14, 2012.

Recommend Tool for Future IW Use?

According to a midlevel operations research analyst with experience using the IWTWG, the most effective M&S tools are wargames that draw organizational and materiel comparisons in IW environments.[27] In other words, they should be able to show meaningful correlation across different variables. If decisionmakers assume that a quantitative model can accurately depict the complex IW environment, then they need to be comfortable with the metrics—or in the case of simulation, variable sets that they recommend and approve. The IWTWG provides commanders with information regarding the population's perception of their campaigns, thus providing essential information to consider when determining what kind of effects their operations will have on the civilian population. However, without the results of the model validation process, it is difficult to determine whether this tool will be useful for future IW conflicts. It is also necessary to keep in mind that no model or simulation can replicate reality with absolute accuracy, nor is it possible to describe to a commander the degree to which any model, simulation, or simulation outcome differs from reality. Therefore, simulation tools like the IWTWG are most effective when used to inform general approaches to IW/COIN, and probably least effective in providing direct operational decision support.

Additional Models and Simulations

Throughout the defense analytic community, a variety of models, wargames, and simulations exist to provide analysts with ways of addressing commanders' decisions. This section provides a sample of these models, simulations, and tools. It is by no means complete—there are many individual IW/COIN-related models, simulations, and tools either in use or in development at various levels across DoD and the Intelligence Community.[28]

Agent-Based Rational Choice Stakeholder Model

The Agent-Based Rational Choice Stakeholder (ABRCS) is a suite of models that includes LODESTONE and Senturion. ABRCS models are designed to predict outcomes of complex political issues and simulate the interactions among stakeholders that lead to these outcomes. LODESTONE is a RAND-developed version of ABRCS that relies on the axioms of spatial politics, strategic game theory, expected utility theory, and rational choice theory to forecast complex decisionmaking outcomes involving

---

[27] Interview with analyst.

[28] These would include any model, simulation, or tool designed to support full-spectrum operations that would also include IW. For example, an embarkation model used by the Marine Corps to help it load ships for deployment to any operation, including OIF or OEF, would be included in this list. Considering all military tasks circumscribed in the four categories we identify, the possible range of models, simulations, and tools is vast.

two or more stakeholders. The origin of these models is in Defense Advanced Research Projects–sponsored research and development conducted in the 1980s. As described in publications of the time, the 1980s versions of the model used a proxy for estimating utility based on the correlation between the rank ordering of stakeholders' preferences across alternative positions; a 1997 version of the model used a more theoretically grounded approach to estimating utility as a function of the distance between stakeholders' positions.[29] In the 2010s, LODESTONE has been used to estimate stakeholder opinions regarding Afghan support for coalition forces in Afghanistan and opinions about the success of U.S. troop surges in Afghanistan.[30]

LODESTONE has three basic capabilities. First, based on the stakeholders' ability to influence the outcome, the model produces a static forecast using weighted spatial voting models to identify the most likely outcome. Second, the model "calculates the expected utility gains or losses each stakeholder might have from challenging or making a proposal to the other stakeholder." Finally, the bargaining continues until the model creates a final prediction of the outcome or until a stakeholder stops the bargaining.[31]

Senturion is a predictive analysis software simulation from the National Defense University (NDU). NDU's Center for Technology and National Security Policy has been testing Senturion since 2002 and began implementing it for real-life DoD challenges shortly thereafter. Specifically, Senturion "analyzes the political dynamics within local, domestic, and international contexts and predicts how the policy positions of competing interests will evolve over time."[32] Using agent-based modeling as a foundation,[33] underpinned by microeconomic, decisionmaking, and political science theory, Senturion analyzes the behavior of individuals and groups who influence political outcomes (the "agents") through algorithms rather than statistics or probability.[34] This approach yields output that allows users to predict future activity based on observation and analysis of past behavior.

---

[29] This paragraph was adapted from RAND work on forecasting from 2011. For a discussion of this family of models, see Eric Larson, Richard E. Darilek, Daniel Gibran, Brian Nichiporuk, Amy Richardson, Lowell H. Schwartz, and Cathryn Quantic Thurston, *Foundations of Effective Influence Operations: A Framework for Enhancing Army Capabilities,* Santa Monica, Calif.: RAND Corporation, MG-654-A, 2009, pp. 46–51; "Assessment of Expected Utility Modeling for Influence Operations," in Larson, Darilek, Kaye, Morgan, Nichiporuk, Dunham-Scott, Thurston, and Leuschner, 2009, pp. 107–117; and Eric V. Larson, Derek Eaton, Brian Nichiporuk, and Thomas S. Szayna, *Assessing Irregular Warfare: A Framework for Intelligence Analysis,* Santa Monica, Calif., MG-668-A, 2008, pp. 29–31.

[30] This section is adapted from RAND work on forecasting from 2011.

[31] This section is adapted from RAND work on forecasting from 2011.

[32] Abdollahian et al., 2006, p. 1.

[33] According to Abdollahian et al. (2006), agent-based modeling "applies a set of mathematical algorithms against rules that structure a simulation of the behavior of 'agents,'" p. 1.

[34] Abdollahian et al., 2006, p. 1.

Senturion uses the insights of SMEs to identify and define the stakeholders, their positions, and their willingness to advocate for policy positions. This process is conducted systematically through the agent-based approach to yield less-biased, more-accurate analysis of the political landscape.[35] NDU used Senturion to simulate interactions among stakeholders in Iraq's stability after the 2003 invasion by analyzing the consequences of regime change on stability in Iraq.[36] SMEs from the RAND Corporation, Brookings Institution, and NDU's own Institute for National Strategic Studies provided the data. Analysis began in 2002 and continued through subsequent iterations of the simulation. This process was repeated—generally with success—for more than 18 months.[37] The software was accurate in predicting the policy outcomes of the U.S. military campaign for both U.S. and Iraqi stakeholders. The Defense Intelligence Agency and the Office of the Secretary of Defense (OSD) judged it to be successful for its accurate predictions of individual stakeholder behavior once OIF began and as events unfolded, as well as its ability to predict the timing of unexpected defections and alliances.[38]

LODESTONE and Senturion's foundation in social science and decisionmaking theory, as well as their agent-based modeling approach, make them potentially useful in analyzing how stakeholders will respond to aspects of an IW campaign. While these models supported some analysis of major combat operations for OIF, we did not have sufficient information to suggest this success would be replicable in other campaigns because they are future planning tools, dependent upon the accuracy of the predictions that inform them.

Cultural Geography Model

TRAC-Monterey developed the Cultural Geography Model as a way to understand civilian responses to IW activity. The model, which is part of the IWTWG, is agent-based and uses survey-type questions about security and stability-related issues to determine the population's response to Blue force IW action. The result, according to TRAC-Monterey, is similar to polling data. The Cultural Geography Model is underpinned by social science and human behavior theories to account for the multiple agents, objects, and laws inherent in an IW environment. Within this multi-agent system, "agents" are capable of acting with "objects" in their area of operations and "laws" exist within the model to govern these interactions.[39]

---

[35] Abdollahian et al., 2006, p. 2.

[36] See Michael Baranick, Mark Abdollahian, Brian Efird, and Jacek Kugler, "Stability and Regime Change in Iraq: An Agent Based Modeling Approach," presentation to Military Operations Research Society, Summer 2004.

[37] Abdollahian et al., 2006, p. 7.

[38] Abdollahian et al., 2006, p. 9.

[39] Jonathan Alt, Stephen Lieberman, and Thomas Anderson, "The Cultural Geography Model: Evaluating the Impact of Tactical Operational Outcomes on a Civilian Population in an Irregular Warfare Environment," *Journal of Defense Modeling and Simulation*, Vol. 6, No. 4, 2009, p. 185.

Preparing data for the Cultural Geography Model is similar to developing a military intelligence preparation of the environment. Relevant data that would affect an operation are fed into the intelligence assessment as they are received, beginning with sociodemographic information from available open-source data and from social scientists with insight into the region. Data regarding the issues and needs that are important to the population, as well information about their history and beliefs, are then inputted, followed by information regarding the threat group (history, beliefs, motivations, goals) and their relationship to, and potential influence over, the local population.[40] The model takes a Bayesian Belief Network approach to analyzing these data, which predicts future actions by relying on past experience. This approach was chosen for its relative simplicity, flexibility, and transparency.[41] The narrative paradigm is also used to gather data. This theory posits that people are storytellers who make decisions based on history, their culture, and basic assumptions about other people in their communities.[42] During the course of the analysis, these data are plotted against tactical operations to determine any cause-and-effect relationships these operations have on the population. During TRAC-Monterey's case study test of the model, regression analysis was also performed to examine what the overall response of the population was to Blue force action. This method was deemed replicable for a variety of demographics to provide commanders information regarding the effects of their operations on the population's main stability concerns.[43]

The Cultural Geography Model is a complex model rooted in behavioral and social science theory. As such, it requires training investments to ensure that those using it are adept at both the theories that serve as its foundation and the methods by which its analysis is conducted. The model is data intensive and the validity of these data rests largely on the abilities of experts, such as anthropologists, to confirm their accuracy. However, the type of information collected—opinions, historical narratives—could be inaccurate, as it is dependent upon a person's correct memory of events. Verifying these memories is not always possible. Further, little data exist regarding the model's practical application to IW scenarios. It has the potential to provide commanders with innovative insight into how the population perceives their actions, but as with all recently developed IW models and simulations, it will require extended application for validation.

---

[40] Alt, Lieberman, and Anderson, 2009, pp. 186–187.

[41] Alt, Lieberman, and Anderson, 2009, p. 188.

[42] Lisa Jean Blair and Richard F. Brown, "Cultural Geography Model Validation," *Proceedings of the 19th Conference on Behavior Representation in Modeling and Simulation*, Charleston, S.C., March 21–24, 2010, pp. 91–92.

[43] Alt, Lieberman, and Anderson, 2009, p. 197.

Analysis and Conclusions

Although some of the wargames, models, and simulations discussed in this appendix focus specifically on stability, each strives in some way to answer the same central question: How do IW operations affect the civilian population, and how can commanders shape operations to improve outcomes? M&S tools have the ability to answer this complex question and many other questions like it, but they require good data, a sound theoretical framework, properly trained operators, and expert participants. The expense that comes with this investment, not to mention the time it takes for organizations to devote personnel and resources to the tool, is not always sustainable for long enough to have an effect on commanders' decisionmaking. Commander preference must also factor into whether results from models and simulations will be used, as some commanders—and some analysts—are inherently skeptical.

If one accepts—as we suggest in Chapter One—that a simulation is "the manipulation of a model (usually using a computer) in such a way that it operates on time or space to gain some understanding of how the modeled system operates," then a simulation is most effective at helping commanders and analysts understand how certain actions—say, the application of development funds—*might* affect a change in popular sentiment in a notional scenario. It seems appropriate to employ simulations like the PSOM as educational tools or preoperational planning tools. As long as participants understand that the model underlying the simulation is a helpful concept and not a precise, perfect reflection of reality—and that the data are necessarily incomplete and inaccurate to some degree—then they should benefit.

However, there is danger in using a model or simulation as complex as the PSOM as a decision-support tool for an ongoing operation. While a well-informed, knowledgeable commander who has experience with M&S may be able to place simulation results in appropriate context, a less-informed commander might be misled into believing that both the model and the associated simulation were intended as precise, definitive representations of ground truth. Modelers might present caveats to the contrary, but caveats are not always embraced by decisionmakers facing a complex challenge; commanders tend to want clear answers.[44] If a simulation appears to present a clear and useful answer to a complex question, a commander might be tempted to leverage simulation results to make operational decisions. Applying campaign-level M&S as a decision support tool exceeds the intent of most models and simulations, and it exceeds the bounds of practicality: no model or simulation could hope to completely or accurately account for all of the relevant variables associated with human decisionmaking at this level of consideration.

Table A.1 summarizes the models and simulations described in this chapter and links them to intended audiences.

---

[44] This statement is based on the interviews conducted for this research, along with other interviews conducted by RAND analysts with commanders and analysts in Afghanistan and supporting decisions in Afghanistan between 2009 and 2012; Connable, 2012.

Table A.1
Summary of Models, Simulations, and Wargames Used in IW Analysis

Model	Creator	OIF or OEF	Questions Answered	Analysts' Impressions	Results Used? By Whom?	Recommend for Future IW Use?
PSOM	Dstl	Both	Popular support of IW activities; future planning	Concerns about where data come from; generally favorable, if slightly uninformed, perceptions of its utility	Commanders in OIF and OEF theaters	Yes—within its constraints as a future planning tool, not as a predictive model
ATHENA	JPL	Both	Simulate the interaction of civilian population with armed forces	Operators require constant feedback to validate the model; rely on SMEs to validate the data that feed the model	OSD Strategic Multilayered Assessment Office	Results of two real-life applications are classified—lack of access creates difficulty in recommending use for a broader set of IW issues
IW Tactical Wargame	TRAC	Both	Sociocultural factors at play during IW conflict	Mostly favorable	Unable to be determined	Yes—but only to inform general approaches to IW/COIN and not to provide direct operational decision support.
Tactical Conflict Assessment and Planning Framework / District Stability Framework	U.S. Agency for International Development/ Office of Military Affairs	Both	Population-specific concerns regarding stability	Mixed; some thought it too complicated, others thought it fit the bill	U.S. Marine Corps, Army	Yes—with continued validation and methodological improvements
Cultural Geography Model	TRAC-Monterey	N/A	Population perceptions of IW activities	N/A	N/A	Heavily based in theory; needs more practical application to ensure its theories provide useful, accurate data to commanders
Senturion	NDU	OIF, via case studies	Identify and define the positions of IW stakeholders based on observation and analysis of past behavior	N/A	Results of case studies were validated by the Defense Intelligence Agency and OSD	Though analysis conducted with Senturion was commissioned by commanders in both Iraq and Afghanistan, results were deemed proprietary by government analysts

Bibliography

This section is written with two purposes in mind: it is intended to both identify those sources used to inform the research for this research project and provide the reader with a broader source of references on assessment in IW and associated issues. Historical documents in the Vietnam Center and Archive online database hosted by the Texas Tech University can be accessed via http://www.virtual.vietnam.ttu.edu using the search function. Documents in the National Archives online research catalog can be accessed via http://arcweb.archives.gov using the search function.

Abbaszadeh, Nima, Mark Crow, Marianne El-Khoury, Jonathan Gandomi, David Kuwayama, Christopher MacPherson, Meghan Nutting, Nealin Parker, and Taya Weiss, *Provincial Reconstruction Teams: Lessons and Recommendations*, Princeton, N.J.: Woodrow Wilson School of Public and International Affairs, January 2008.

Abdollahian, Mark, et al., *Senturion: A Predictive Political Simulation Model*, Center for Technology and National Security Policy, National Defense University, July 2006.

Abella, Alex, *Soldiers of Reason: The RAND Corporation and the Rise of the American Empire*, Orlando, Fla.: Harcourt, 2008.

Abramowitz, Michael, "Congress, White House Battle Over Iraq Assessment," *Washington Post*, September 10, 2007.

Ackoff, Russell L., "Towards a System of Systems Concepts," *Management Science*, Vol. 17, No. 11, July 1971. As of June 30, 2011:
http://ackoffcenter.blogs.com/ackoff_center_weblog/files/AckoffSystemOfSystems.pdf

———, "From Data to Wisdom," *Journal of Applied Systems Analysis*, Vol. 16, 1989.

———, "Systems Thinking and Thinking Systems," *System Dynamics Review*, Vol. 10, Nos. 2–3, Summer-Fall 1994.

Ahern, Thomas L., Jr., *Vietnam Declassified: The CIA and Counterinsurgency*, Lexington, Ky.: University of Kentucky Press, 2010.

Akst, George, "Analysis for Non-Traditional Security Challenges: Methods and Tools," presentation at Military Operations Research Society Workshop: Analysis for Non-Traditional Security Challenges, Methods and Tools, JHU/APL, Laurel, Md., February 21–23, 2006.

Alberts, David S., John J. Garstka, and Frederick P. Stein, *Network Centric Warfare: Developing and Leveraging Information Superiority*, 2nd ed., Command Control Research Program Publications, 1999.

Allen, Thomas, James Bexfield, and Stuart Starr, "Perspectives on the Analysis Modeling and Simulation Plan," presentation at 13th International Command and Control Research and Technology Symposium, Seattle, Wash., June 17–19, 2008.

Allison, Paul David, *Missing Data*, Thousand Oaks, Calif.: Sage Publications, 2002.

Alt, Jonathan, Stephen Lieberman, and Thomas Anderson, "The Cultural Geography Model: Evaluating the Impact of Tactical Operational Outcomes on a Civilian Population in an Irregular Warfare Environment," *Journal of Defense Modeling and Simulation*, Vol. 6, No. 4, 2009.

Anderson, D., D. Henderson, A. Hummel, R. Spivey, and J. Wray, *Intra-Theater Air Lift Planning—Redux*, TRAC-L-TR-10-044, July 2010.

Anderson, David A., and Andrew Wallen, "Preparing for Economics in Stability Operations," *Military Review*, Vol. 88, No. 2, March-April 2008.

Anderson, Edward G., Jr., "A Dynamic Model of Counterinsurgency Policy Including the Effects of Intelligence, Public Security, Popular Support, and Insurgent Experience," *System Dynamics Review*, Vol. 27, No. 2, April-June 2011.

Anderson, Joseph, "Military Operational Measures of Effectiveness for Peacekeeping Operations," *Military Review*, Vol. 81, No. 5, September-October 2001.

Appleget, Jeff, "PSOM Overview and Peacekeeping Operations Assessment Using PSOM," Briefing delivered to the Naval Postgraduate School, Monterey, Calif., October 2011.

Armstrong, Nicholas J., and Jacqueline Chura-Beaver, "Harnessing Post-Conflict Transitions: A Conceptual Primer," Peacekeeping and Stability Operations Institute, September 2010.

Army Logistics University, *Operations Research/Systems Analysis (ORSA) Fundamental Principles, Techniques, and Applications*, ORSA Committee, October 2011. As of October 31, 2012: http://www.almc.army.mil/ALU_DOCS/ORSABook.pdf

"Army Plenary Brief," presentation at Military Operations Research Society Workshop: Analysis for Non-Traditional Security Challenges: Methods and Tools, JHU/APL, Laurel, Md., February 21, 2006.

Arnold, Matthew, "Human Terrain Mapping in Kapisa Province," *Small Wars Journal*, April 12, 2010.

Arquilla, John, and David Ronfeldt, *In Athena's Camp: Preparing for Conflict in the Information Age*, Santa Monica, Calif.: RAND Corporation, MR-880-OSD/RC, 1997. As of June 30, 2011: http://www.rand.org/pubs/monograph_reports/MR880.html

Arthur, James F., Testimony before the House Armed Services Committee on hamlet pacification in Vietnam, March 1970. As of January 27, 2011: http://www.virtual.vietnam.ttu.edu

Asia Foundation, Afghanistan in 2011: A Survey of the Afghan People, web page, undated. As of August 15, 2012: http://asiafoundation.org/country/afghanistan/2011-poll.php

BACM Research, *Vietnam War Document Archive* (Disk 1 and Disk 2), 2009.

Baker, Pauline H., "Measuring Progress in COIN," presentation at Improving Cooperation Among Nations in Irregular Warfare Analysis, Naval Postgraduate School, Monterey, Calif., December 11–13, 2007.

Baker, Ralph O., "The Decisive Weapon: A Brigade Combat Team Commander's Perspective on Information Operations," *Military Review*, May-June 2006.

———, "HUMINT-Centric Operations: Developing Actionable Intelligence in the Urban Counterinsurgency Era," *Military Review*, Vol. 87, No. 2, 2007.

Baranick, Michael, Mark Abdollahian, Brian Efird, and Jacek Kugler, "Stability and Regime Change in Iraq: An Agent Based Modeling Approach," presentation to Military Operations Research Society, Summer 2004.

Baranick, Michael, David Knudson, Dave Pendergraft, and Paul Evangelista, "Improving Analytical Support to the Warfighter: Campaign Assessments, Operational Analysis, and Data Management, Working Group 1: Data and Knowledge Management," briefing, Military Operations Research Society, April 19–22, 2010. As of September 15, 2010: http://www.mors.org/UserFiles/file/2010%20IW/MORS%20IW%20WG1.pdf

Barber, A. H., "Navy Non-Traditional Modeling and Analysis Challenges," presentation at MORS Workshop: Analysis for Non-Traditional Security Challenges: Methods and Tools, JHU/APL, Laurel, Md., February 21–23, 2006.

Barno, David W., and Andrew M. Exum, *Responsible Transition: Securing U.S. Interests in Afghanistan Beyond 2014*, Center for New American Security, December 7, 2010.

Bar-Yam, Yaneer, *Complexity of Military Conflict: Multiscale Complex Systems Analysis of Littoral Warfare*, New England Complex Systems Institute, April 21, 2003. As of June 30, 2011: http://necsi.edu/projects/yaneer/SSG_NECSI_3_Litt.pdf

Bennett, William, "Media Influencing Modeling in Support of Conflict Modeling, Planning, and Outcome Experimentation (COMPOEX)," presentation at Military Operations Research Society Conference: Irregular Warfare II, MacDill Air Force Base, Fla., February 4–6, 2009.

Bernard, H. Russell, *Research Methods in Anthropology: Qualitative and Quantitative Approaches*, 4th ed., Lanham, Md.: AltaMira Press, 2006.

Beyerchen, Alan, "Clausewitz, Nonlinearity, and the Unpredictability of War," *International Security*, Vol. 17, No. 3, Winter 1992–1993.

Biddle, Stephen, *Military Power: Explaining Victory and Defeat in Modern Battle*, Princeton, N.J.: Princeton University Press, 2004.

Birdwell, Trey, and John A. Klemunes, "Tools of War," *Engineer*, January-March 2004.

Birtle, Andrew J., *U.S. Army Counterinsurgency and Contingency Operations Doctrine, 1860–1941*, Washington, D.C.: U.S. Army Center of Military History, 2004.

———, *U.S. Army Counterinsurgency and Contingency Operations Doctrine, 1942–1976*, Washington, DC: U.S. Army Center of Military History, 2006.

Bitinas, Edmund J., "Pythagoras Overview," presentation at Improving Cooperation Among Nations in Irregular Warfare Analysis, Naval Postgraduate School, Monterey, Calif., December 11–13, 2007.

Blair, Lisa Jean, and Richard F. Brown, "Cultural Geography Model Validation," *Proceedings of the 19th Conference on Behavior Representation in Modeling and Simulation*, Charleston, S.C., March 21–24, 2010.

Blalock, Hubert M., Jr., ed., *Measurement in the Social Sciences: Theories and Strategies*, Chicago, Ill.: Aldine Publishing Company, 1974.

Body, Howard, and Colin Marston, "The Peace Support Operations Model: Origins, Development, Philosophy and Use," *Journal of Defense Modeling and Simulation: Applications, Methodology, Technology*, Vol. 8, No. 2, 2010.

Bonder, Seth, "Army Operations Research—Historical Perspectives and Lessons Learned," *Operations Research*, January-February 2002.

Boot, Max, *War Made New: Weapons, Warriors, and the Making of the Modern World*, New York: Penguin Group, 2006.

Borkowski, Piotr, and Jan Mielniczuk, "Postmodel Selection Estimators of Variance Function for Nonlinear Autoregression," *Journal of Time Series Analysis*, Vol. 31, No. 1, January 2010, pp. 50–63.

Bousquet, Antoine, *The Scientific Way of Warfare: Order and Chaos on the Battlefields of Modernity*, New York: Columbia University Press, 2009.

Bowman, Christopher W., *Operational Assessment—The Achilles Heel of Effects-Based Operations?* thesis, Newport, R.I.: Naval War College, May 13, 2002.

Bracken, Jerome, Moshe Kress, and Richard E. Rosenthal, eds., *Warfare Modeling*, Danvers, Mass.: John Wiley and Sons, 1995.

Brannen, Kate, "Combat Brigades in Iraq Under Different Name," *Army Times*, August 19, 2010. As of August 24, 2012:
http://www.armytimes.com/news/2010/08/dn-brigades-stay-under-different-name-081910/

Brigham, Erwin R., "Pacification Measurement in Vietnam: The Hamlet Evaluation System," presentation at SEATO Internal Security Seminar, Manila, Philippines, June 3–10, 1968. As of June 30, 2011:
http://www.cgsc.edu/carl/docrepository/PacificationVietnam.pdf

———, "Pacification Measurement," *Military Review*, May 1970.

Brookings Institution, Saban Center for Middle East Policy, *Iraq Index: Tracking Reconstruction and Security in Post-Saddam Iraq*, Washington, D.C., last updated August 24, 2010. See archive of past reports. As of June 30, 2011:
http://www.brookings.edu/saban/iraq-index.aspx

Brown, Jason M., "To Bomb or Not to Bomb? Counterinsurgency, Airpower, and Dynamic Targeting," *Air & Space Power Journal*, Vol. 21, No. 4, Winter 2007.

Brown, James, Erik W. Goepner, and James M. Clark, "Detention Operations, Behavior Modification, and Counterinsurgency," *Military Review*, May-June 2009.

Bryman, Alan, Saul Becker, and Joe Sempik, "Quality Criteria for Quantitative, Qualitative and Mixed Methods Research: A View from Social Policy," *Social Research Methodology*, Vol. 11, No. 4, October 2008.

Bunker, Ellsworth, "October 2–15, 1968: The Breakthrough in Paris," official telegram from the U.S. Embassy in Saigon, Vietnam, to U.S. Department of State, Washington, D.C., October 2, 1968a, (BACM Research, Disk 2).

———, "Telegram from the Embassy in Vietnam to the Department of State," official telegram from the U.S. Embassy in Saigon, Vietnam, to the U.S. Department of State, Washington, D.C., October 19, 1968b (BACM Research, Disk 2).

Burton, Janice, "Army Executive Irregular Warfare Conference Charts Army's Path," *Special Warfare*, Vol. 22, No. 6, November-December 2009.

CAA—*See* Center for Army Analysis.

Cabayan, Hriar, *Rich Contextual Understanding of Pakistan and Afghanistan (PAKAF): A Strategic Multi-Layer Assessment*, unpublished workshop report, April 12, 2010a.

———, *Subject Matter Analysis PAKAF Rich Contextual Understanding, MG Flynn Final Progress Report*, unpublished workshop report, October 25, 2010b.

Cale, Paul P., "The United States Military Advisory Group in El Salvador, 1979–1992," thesis, U.S. Army Command and General Staff College, 1996.

Callwell, C. E., *Small Wars: Their Principles and Practice*, 3rd Edition, Lincoln, Neb.: University of Nebraska Press, 1996 (Original Publication 1896).

Campbell, Jason, Michael E. O'Hanlon, and Jeremy Shapiro, *Assessing Counterinsurgency and Stabilization Missions*, Washington, D.C.: Brookings Institution, policy paper 14, May 2009a. As of June 30, 2011:
http://www.brookings.edu/~/media/Files/rc/papers/2009/05_counterinsurgency_ohanlon/05_counterinsurgency_ohanlon.pdf

———, "How to Measure the War," *Policy Review*, No. 157, October-November 2009b. As of June 30, 2011:
http://www.hoover.org/publications/policy-review/article/5490

Carmines, Edward G., and Richard A. Zeller, *Reliability and Validity Assessment*, Beverly Hills, Calif.: Sage Publications, 1979.

Carpenter, P. Mason, and William F. Andrews, "Effects-Based Operations: Combat Proven," *Joint Force Quarterly*, No. 52, First Quarter 2009.

Cebrowski, Arthur K., and John J. Garstka, "Network-Centric Warfare: Its Origin and Future," *Proceedings*, January 1998.

Center for Army Analysis, "Operation OUTREACH: Tactics, Techniques, and Procedures," No. 03-27, October 2003.

———, "ORSA Handbook for the Senior Commander," March 1, 2008.

———, "Assessment and Measures of Effectiveness in Stability Ops," Handbook, Vol. 10-41, May 2010a.

———, "Afghanistan Consolidated Knowledge System," briefing, August 30, 2010b.

———, "Afghanistan Provincial Reconstruction Team," Handbook 11-16, February 2011. As of November 15, 2012:
http://usacac.army.mil/cac2/call/docs/11-16/11-16.pdf

———, *Analytic Support to Combat Operations in Iraq (2002–2011)*, Deployed Analyst History Report, Vol. 1, March 2012.

Chamberlain, Robert G., and William H. Duquette, "Athena in 2011," paper presented at the Military Operations Research Society Special Meeting on Operations Research Methods in Support of Countering Transnational Threats in McLean, Va., December 12–14, 2011.

CIA—*See* U.S. Central Intelligence Agency.

Cioppa, Thomas, email exchange with the author, October 8, 2010.

Cioppa, Thomas, Loren Eggen, and Paul Works, "Improving Analytical Support to the Warfighter: Campaign Assessments, Operational Analysis, and Data Management, Working Group 4: Current Ops Analysis—Tactical," briefing, Military Operations Research Society, April 19–22, 2010. As of June 30, 2011:
http://www.mors.org/UserFiles/file/2010%20IW/MORS%20IW%20WG4.pdf

Citizen, Jesse, "Management of DoD Modeling and Simulation," briefing, DoD Modeling and Simulation Coordination Office, undated.

Claflin, Bobby, Dave Sanders, and Greg Boylan, "Improving Analytical Support to the Warfighter: Campaign Assessments, Operational Analysis, and Data Management, Working Group 2: Campaign Assessments," working group briefing, Military Operations Research Society conference, April 19–22, 2010. As of September 15, 2010:
http://www.mors.org/UserFiles/file/2010%20IW/MORS%20IW%20WG2.pdf

Clancy, James, and Chuck Crossett, "Measuring Effectiveness in Irregular Warfare," *Parameters*, Summer 2007.

Clark, Clinton R., and Timothy J. Cook, "A Practical Approach to Effects-Based Operational Assessment," *Air and Space Power Journal*, Vol. 22, No. 2, Summer 2008.

Clark, Dorothy K., and Charles R. Wyman, *An Exploratory Analysis of the Reporting, Measuring, and Evaluating of Revolutionary Development in South Vietnam*, McLean, Va.: Research Analysis Corporation, 1967.

Colby, William E., and Peter Forbath, *Honorable Men: My Life in the CIA*, New York: Simon and Schuster, 1978.

Coleman, Edward J., and Rico R. Bussey, "A Primer on Indirect Fire Crater Analysis in Iraq and Afghanistan," *Field Artillery Journal*, July-August 2005.

Commission to Investigate the Lebanon Campaign in 2006, final report, January 30, 2008. As of June 30, 2011:
http://www.mfa.gov.il/MFA/MFAArchive/2000_2009/2008/Winograd%20Committee%20submits%20final%20report%2030-Jan-2008

Connable, Ben, "All Our Eggs in a Broken Basket: How the Human Terrain System Is Undermining Sustainable Military Cultural Competence," *Military Review*, Vol. 89, No. 2, March-April 2009.

———, "The Massacre That Wasn't," in *U.S. Marines in Iraq, 2004–2008: Anthology and Annotated Bibliography*, Washington, D.C.: History Division, U.S. Marine Corps, 2010.

———, *Embracing the Fog of War: Assessment and Metrics in Counterinsurgency*, Santa Monica, Calif.: RAND Corporation, MG-1086-DOD, 2012. As of August 3, 2013:
http://www.rand.org/pubs/monographs/MG1086.html

Connable, Ben, and Martin C. Libicki, *How Insurgencies End*, Santa Monica, Calif.: RAND Corporation, MG-965-MCIA, 2010. As of June 30, 2011:
http://www.rand.org/pubs/monographs/MG965.html

Conway, J. Edward, "Analysis in Combat: The Deployed Threat Finance Analyst," *Small Wars Journal*, July 5, 2012.

Cooley, Tim, "Current Trends in M&S ROI Calculation: An Addendum to 'Calculating ROI Investment for US DOD M&S,'" *M&S Journal*, Fall 2012.

Cooper, Chester L., Judith E. Corson, Laurence J. Legere, David E. Lockwood, and Donald M. Weller, *The American Experience with Pacification in Vietnam*, Vol. 1: *An Overview of Pacification*, Institute for Defense Analysis, March 1972a.

———, *The American Experience with Pacification in Vietnam*, Vol. 2: *Elements of Pacification*, Institute for Defense Analysis, March 1972b.

Cordesman, Anthony H., *The Uncertain "Metrics" of Afghanistan (and Iraq)*, Washington, D.C.: Center for Strategic and International Studies, May 18, 2007. As of June 30, 2011:
http://www.csis.org/media/csis/pubs/070521_uncertainmetrics_afghan.pdf

———, *Analyzing the Afghan-Pakistan War*, Washington, D.C.: Center for Strategic and International Studies, draft, July 28, 2008. As of September 15, 2010:
http://csis.org/files/media/csis/pubs/080728_afghan_analysis.pdf

———, "The Afghan War: Metrics, Narratives, and Winning the War," briefing, Center for Strategic and International Studies, June 1, 2010.

Corson, William R., *The Betrayal*, New York: W. W. Norton and Company, 1968.

Cosmos, Graham A., *MACV: The Joint Command in the Years of Withdrawal, 1968–1973*, Washington, D.C.: U.S. Army Center of Military History, 2006. As of June 30, 2011:
http://www.history.army.mil/html/books/091/91-7/CMH_Pub_91-7.pdf

Crowell, Benjamin, *Light and Matter*, Creative Commons Attributions, 2011. As of May 22, 2012:
http://www.lightandmatter.com/lm/

Cushman, John H., *Senior Officer Debriefing Report of Major General John H. Cushman, RCS CSFOR-74*, January 14, 1972. As of June 30, 2011:
http://www.dtic.mil/cgi-bin/GetTRDoc?Location=U2&doc=GetTRDoc.pdf&AD=AD0523904

Daddis, Gregory A., *No Sure Victory: Measuring U.S. Army Effectiveness and Progress in the Vietnam War*, New York: Oxford University Press, 2011.

Darilek, Richard E., Walter L. Perry, Jerome Bracken, John Gordon IV, and Brian Nichiporuk, *Measures of Effectiveness for the Information-Age Army*, Santa Monica, Calif.: RAND Corporation, MR-1155-A, 2001. As of June 30, 2011:
http://www.rand.org/pubs/monograph_reports/MR1155.html

DataCards, homepage, undated. As of August 3, 2013 (logon required):
https://www.datacards.org

Davis, Paul K., *Effects-Based Operations (EBO): A Grand Challenge for the Analytical Community*, Santa Monica, Calif.: RAND Corporation, MR-1477-USJFCOM/AF, 2001. As of June 30, 2011:
http://www.rand.org/pubs/monograph_reports/MR1477.html

Davis, Paul K., Kim Cragin, eds., *Social Science for Counterterrorism: Putting the Pieces Together*, Santa Monica, Calif.: RAND Corporation, MG-849-OSD, 2009. As of August 3, 2013:
http://www.rand.org/pubs/monographs/MG849.html

Dean, Arleigh William, "Fighting Networks: The Defining Challenge of Irregular Warfare," thesis, Monterey, Calif.: Naval Postgraduate School, June 2011.

Defense Science Board, *Strategic Communication*, Washington, D.C.: Office of the Under Secretary of Defense for Acquisition, Technology, and Logistics, September 2004a.

———, *Summer Study on Transition to and from Hostilities*, Washington, D.C.: Office of the Under Secretary of Defense for Acquisition, Technology, and Logistics, December 2004b.

———, *Understanding Human Dynamics*, Washington, D.C.: Office of the Under Secretary for Acquisition, Technology, and Logistics, March 2007.

———, *Fulfillment of Urgent Operational Needs*, Washington, D.C.: Office of the Under Secretary of Defense for Acquisition, Technology, and Logistics, July 2009b.

———, *Counterinsurgency (COIN) Intelligence, Surveillance, and Reconnaissance (ISR) Operations*, Washington, D.C., Office of the Under Secretary of Defense for Acquisition, Technology, and Logistics, February 2011.

Demarest, Geoffrey B., and Lester W. Grau, "Maginot Line or Fort Apache? Using Forts to Shape the Counterinsurgency Battlefield," *Military Review*, Vol. 85, No. 6, November-December 2005.

DoD—*See* U.S. Department of Defense.

Donovan, David, *Once a Warrior King: Memories of an Officer in Vietnam*, New York: McGraw-Hill, 1985.

Downes-Martin, Stephen, U.S. Naval War College, interview with the author, Kabul, Afghanistan, May 5, 2010a.

———, "Assessment Process for RC(SW)," unpublished draft, Newport, R.I.: U.S. Naval War College, May 24, 2010b.

———, "Operations Assessment in Afghanistan Is Broken: What Is to Be Done?" *Naval War College Review*, Newport, R.I.: Naval War College Press, Vol. 64, No. 4, 2011.

Dror, Yehezkel, *Some Normative Implications of a Systems View of Policymaking*, Santa Monica, Calif.: RAND Corporation, P-3991-1, 1969. As of June 30, 2011: http://www.rand.org/pubs/papers/P3991-1.html

Duong, Deborah, "Gaps, Tools, and Evaluation Methodologies for Analyzing the Global War on Terror," presentation at Improving Cooperation Among Nations in Irregular Warfare Analysis, Naval Postgraduate School, Monterey, Calif., December 11–13, 2007.

———, "Agent Based Simulation," presentation at Military Operations Research Society Conference: Irregular Warfare II, MacDill Air Force Base, Fla., February 4–6, 2009.

———, "Nexus Network Learner," presentation at Military Operations Research Society Conference: Irregular Warfare II, MacDill Air Force Base, Fla., February 4–6, 2009.

Duong, Deborah, Lauren Murphy, and Will Ellerbe, "The Oz Wargame Integration Toolkit: Supporting Wargames for Analysis," presentation at Military Operations Research Society Conference: Irregular Warfare II, MacDill Air Force Base, Fla., February 4–6, 2009.

Duong, Deborah et al., "Nexus: An Interpretative Social Simulation of Corruption," briefing, OSD/PA&E Simulation Analysis Center, Marine Corps Combat Development Command, April 25, 2008.

Duquette, William H., "Athena User's Guide, Athena A&RO Simulation, V3," Jet Propulsion Laboratory, January 2012.

Earnest, David C., "Growing a Virtual Insurgency: Using Massively Parallel Gaming to Simulate Insurgent Behavior," *The Journal of Defense Modeling and Simulation: Applications, Methodology, Technology*, Vol. 6, No. 2, 2009.

Eisenstadt, Michael, and Jeffrey White, "Assessing Iraq's Sunni Arab Insurgency," *Military Review*, Vol. 86, No. 3, May–June 2006.

Eisler, David F., "Counter-IED Strategy in Modern War," *Military Review*, January-February 2012.

Eles, Philip, U.S. Central Command, interview with the author, Tampa, Fla., March 24, 2010.

Emery, Norman, "Information Operations in Iraq," *Military Review*, Vol. 84, No. 3, May-June 2004.

———, "Irregular Warfare Information Operations: Understanding the Role of People, Capabilities, and Effects," *Military Review*, Vol. 88, No. 6, November-December 2008.

Engelhardt, Tom, "Afghanistan by the Numbers: Is the War Worth It? The Cost in Dollars, Years, Public Opinion, and Lives," *Salon.com*, September 14, 2009. As of January 27, 2011: http://www.salon.com/news/feature/2009/09/14/afghanistan

Enthoven, Alain C., and K. Wayne Smith, *How Much Is Enough? Shaping the Defense Program, 1961–1969*, New York: Harper and Row/Santa Monica, Calif.: RAND Corporation, 1971/2005. As of June 30, 2011: http://www.rand.org/pubs/commercial_books/CB403.html

Evans, D. J., "Operational Analysis in Support of HQ ISAF, Kabul Afghanistan, 2002," in *The Cornwallis Group VIII: Analysis for Governance and Stability*, Farnborough, UK: Defence Science and Technology Laboratory, 2002.

Feinberg, Jerry M., "Understanding the Value of M&S," *M&S Journal*, Winter Edition 2011.

Fike, Angela, Joyce Nagle, Bill Goran, Niki Goerger, and Jack Jackson, "Contingency Operations Tiger Team Initiative and Representing Urban Cultural Geography in Stability Operations," presentation at Military Operations Research Society Conference: Irregular Warfare II, MacDill Air Force Base, Fla., February 4–6, 2009.

Fischer, Hannah, *Iraqi Civilian Death Estimates*, Washington, D.C.: Congressional Research Service, RS22537, August 27, 2008.

Flynn, Michael T., Matt Pottinger, and Paul Batchelor, *Fixing Intel: A Blueprint for Making Intelligence Relevant in Afghanistan*, Washington, D.C.: Center for a New American Security, January 2010. As of June 30, 2011:
http://www.cnas.org/files/documents/publications/AfghanIntel_Flynn_Jan2010_code507_voices.pdf

Folker, Robert D. Jr., *Intelligence Analysis in Theater Joint Intelligence Centers: An Experiment in Applying Structured Analytic Methods*, Washington, D.C.: Joint Military Intelligence College, occasional paper No. 7, January 2000.

Forterra Systems, OLIVE (On-Line Interactive Virtual Environment), 2004.

Friedman, Jeffrey A., "Manpower and Counterinsurgency: Empirical Foundations for Theory and Doctrine," *Security Studies*, Vol. 20, No. 4, November 29, 2011.

Gaffney, Helen, and Alasdair Vincent, "Modeling Information Operations in a Tactical-Level Stabilization Environment," *The Journal of Defense Modeling and Simulation: Applications, Methodology, Technology,* Vol. 8, No. 2, 2011.

Gaghan, Frederick, "Attacking the IED Network," briefing, Joint Improvised Explosive Device Organization, May 5, 2011. As of May 30, 2013:
http://www.dtic.mil/ndia/2011GlobalExplosive/Gaghan.pdf

Galula, David, *Counterinsurgency Warfare: Theory and Practice,* Westport, Conn.: Praeger Security International, 1964.

GAMS homepage, undated. As of September 4, 2012:
http://www.gams.com/

Gartner, Scott Sigmund, *Strategic Assessment in War*, New Haven, Conn.: Yale University Press, 1997.

Gates, Robert M., Secretary of Defense, Statement to the U.S. Senate Appropriations Committee for Defense, June 16, 2010. As of September 15, 2010:
http://appropriations.senate.gov/ht-defense.cfm?method=hearings.
download&id=fac53160-5308-4ffd-9b61-bb8112599ce2

Gayvert, David, "Teaching New Dogs Old Tricks: Can the Hamlet Evaluation System Inform the Search for Metrics in Afghanistan?" *Small Wars Journal Online*, September 8, 2010. As of June 30, 2011:
http://smallwarsjournal.com/blog/journal/docs-temp/531-gayvert.pdf

Gelb, Leslie H., and Richard K. Betts, *The Irony of Vietnam: The System Worked*, Washington, D.C.: Brookings Institution Press, 1979.

Gentile, Gian P., "A Strategy of Tactics: Population-centric COIN and the Army," *Parameters*, Autumn 2009.

Gibson, James William, *The Perfect War: Technowar in Vietnam*, New York: Atlantic Monthly Press, 2000.

Gilbert, Nigel, *Agent-Based Models*, Thousand Oaks, Calif.: Sage Publications, 2008.

Goldstein, Gordon M., *Lessons in Disaster: McGeorge Bundy and the Path to War in Vietnam*, New York: Henry Holt and Company, 2008.

Gomez, Jimmy A., "The Targeting Process: D3A and F3EAD," *Small Wars Journal*, July 16, 2011.

Gons, Eric, Jonathan Schroden, Ryan McAlinden, Marcus Gaul, and Bret van Poppel, "Challenges of Measuring Progress in Afghanistan Using Violence Trends: The Effects of Aggregation, Military Operations, Seasonality, Weather, and Other Causal Factors," *Defense and Security Analysis*, Vol. 28, No. 2, June 2012.

Goode, Steven, "A Historical Basis for Force Requirements in Counterinsurgency," *Parameters*, Strategic Studies Institute, Winter, 2009. As of August 8, 2013: http://strategicstudiesinstitute.army.mil/pubs/parameters/Articles/09winter/goode.pdf

Grau, Lester W., "Something Old, Something New: Guerrillas, Terrorists, and Intelligence Analysis," *Military Review*, July-August 2004.

Graves, Greg, Patricia Murphy, and Frederick Cameron, "Improving Analytical Support to the Warfighter: Campaign Assessments, Operations Analysis, and Data Management, Working Group 4: Current Operations Analysis—Strategic and Operational Level," briefing, Military Operations Research Society, April 19–22, 2010. As of June 30, 2011: http://www.mors.org/UserFiles/file/2010%20IW/MORS%20IW%20WG5.pdf

Greenberg, Brandi, and Frank Mullen, "The Modeling and Simulation Catalog for Discovery, Knowledge, and Re-Use," *M&S Journal*, Fall 2011.

Greenwood, T. C., and T. X. Hammes, "War Planning for Wicked Problems," *Armed Forces Journal*, December 2009. As of June 30, 2011 (logon required): http://armedforcesjournal.com/2009/12/4252237/

Gregersen, Hal, and Lee Sailer, "Chaos Theory and Its Implications for Social Science Research," *Human Relations*, Vol. 46, No. 7, July 1993.

Gregor, William J., "Military Planning Systems and Stability Operations," *PRISM*, Vol. 1, No. 3, 2010. As of June 30, 2011: http://www.army.mil/-news/2010/07/22/42647-military-planning-systems-and-stability-operations/index.html

Grier, Cindy, Steve Stephens, Renee Carlucci, Stuart Starr, Cy Staniec, LTC(P) Clark Heidelbaugh, Tim Hope, Don Brock, Gene Visco, and Jim Stevens, "Improving Analytical Support to the Warfighter: Campaign Assessments, Operational Analysis, and Data Management: Synthesis Group," briefing, Military Operations Research Society, April 19–22, 2010. As of June 30, 2011: http://www.mors.org/UserFiles/file/2010%20IW/Synthesis%20Group%20Out%20Brief%20Final.pdf

Grossman, Elaine M., "JFCOM Draft Report Finds U.S. Forces Reverted to Attrition in Iraq," *Inside the Pentagon*, March 25, 2004.

Grubbs, Lee K., "Is There a Deep Fight in a Counterinsurgency?" *Military Review*, Vol. 85, No. 4, July-August 2005.

Haken, Nate, Joelle Burbank, and Pauline H. Baker, "Casting Globally: Using Content Analysis for Conflict Assessment and Forecasting," *Military Operations Research*, Vol. 15, No. 2, 2010, pp. 5–19.

Halberstam, David, *The Best and the Brightest*, New York: Random House, 1972.

Hall, CSM Michael T., and GEN Stanley A. McChrystal, commander, U.S. Forces–Afghanistan/International Security Assistance Force, Afghanistan, "ISAF Commander's Counterinsurgency Guidance," Kabul, Afghanistan: Headquarters, International Security Assistance Force, 2009. As of June 30, 2011:
http://www.nato.int/isaf/docu/official_texts/counterinsurgency_guidance.pdf

Halmos, Paul R., *Measure Theory*, New York: Van Nostrand, 1950.

Hamilton, James D., *Time Series Analysis*, Princeton, N.J.: Princeton University Press, 1994.

Hammes, Thomas X., *The Sling and the Stone: On War in the 21st Century*, St. Paul, Minn.: Zenith Press, 2004.

———, "Countering Evolved Insurgent Networks," *Military Review*, Vol. 86, No. 4, July-August 2006.

Hammond, William M., *The United States Army in Vietnam: Public Affairs The Military and the Media—1962–1968*, Washington, D.C.: U.S. Government Printing Office, 1988.

Hanley, Nathan, and Helen Gaffney, "The Peace Support Operations Model: Modeling Techniques Present and Future," *The Journal of Defense Modeling and Simulation: Applications, Methodology, Technology*, Vol. 8, No. 2, 2011.

Hanley, John T., Jr., "Potential Applicability of Ops Analysis Techniques," presentation at MORS Workshop: Analysis for Non-Traditional Security Challenges, Methods and Tools, JHU/APL, Laurel, Md., February 21–23, 2006.

Hannan, Michael J., "Operational Net Assessment: A Framework for Social Network Analysis," *IOsphere*, Fall 2005, pp. 27–32.

Hartley, Dean S., Lee Lacey, and Paul Works, "IW Ontologies," briefing, INFORMS National Meeting, Charlotte, N.C., November, 2011. As of October 31, 2012:
http://home.comcast.net/~dshartley3/DIMEPMESIIGroup/Documents/INFORMS%20TRAC%20IW%20Ontology_111311.ppt

Harvey, Derek J., director, Afghanistan-Pakistan Center of Excellence, U.S. Central Command, discussion with the author, January 22, 2010.

Headquarters, U.S. Army, *Operations Against Guerrilla Forces*, Field Manual 31–20, Washington, D.C., 1951.

———, *Operations Against Irregular Forces*, Field Manual 31–15, Washington, D.C.: Headquarters, Department of the Army, May 1961.

———, *Counterguerrilla Operations*, Field Manual 31–16, Washington, D.C., February 1963.

———, *Civil Affairs Operation*, Field Manual, 41–1, Washington, D.C., October, 1969.

———, *Intelligence Analysis*, Field Manual 34–3, Washington, D.C., March 1990.

———, *Mission Command: Command and Control of Army Forces*, Field Manual 6–0, Washington, D.C., August 2003.

———, *Intelligence*, Field Manual 2–0, Washington, D.C.: Headquarters, May 2004.

———, *The Army*, Field Manual 1, Washington, D.C., June 2005. As of September 17, 2010:
http://www.army.mil/fm1/

———, *The Operations Process*, Field Manual Interim 5–0.1, Washington, D.C., March 2006a.

———, Field Manual 5–19, Composite Risk Management, Washington, D.C., July 2006b.

———, *Civil Affairs Operations*, Field Manual 3–05.40, Washington, D.C., September 6, 2006c.

————, *Counterinsurgency*, Field Manual 3–24/Marine Corps Warfare Publication 3-33.5, Washington, D.C., December 2006d.

————, *2007 Army Modernization Plan*, Washington, D.C., March 5, 2007.

————, *Operations*, Field Manual 3–0, Washington, D.C., February 27, 2008.

————, *Tactics in Counterinsurgency*, Field Manual 3–24.2, Washington, D.C., April 21, 2009a.

————, *Security Force Assistance*, Field Manual 3–07.1, Washington D.C., May 2009b.

————, *The Operations Process*, Field Manual 5–0, Washington, D.C.: March 2010a.

————, *Army Operating Concept 2016 -2028*, TRADOC Pam 525-3-1, Washington, D.C., August 19, 2010b.

————, "Army Doctrinal Term Changes," unofficial guidelines for Army training developers, spreadsheet, April 29, 2011a.

————, "Army Modified Table of Organization and Equipment (MTOE) Scrub Strategic Issues," Memorandum, Department of the Army Military Operations FMF, December 15, 2011b.

————, *2012 Army Posture Statement, Addendum G—Army Force Generation (ARFORGEN)*, Washington, D.C, 2012. As of November 30, 2012:
https://secureweb2.hqda.pentagon.mil/vdas_armyposturestatement/2012/addenda/addenda_g.aspx

Headquarters, U.S. Marine Corps, *Small Wars Manual*, Washington, D.C., 1940.

————, *Intelligence*, Marine Corps Doctrine Publication 2, Washington, D.C., June 7, 1997a.

————, *Warfighting*, Marine Corps Doctrine Publication 1, Washington, D.C., June 20, 1997b.

————, "A Concept for Distributed Operations," white paper, Commandant of the Marine Corps, April 25, 2005.

Helfstein, Scott, "AQ's Comm: Strategies, Capabilities, and Results," presentation at MORS Conference: Irregular Warfare II, MacDill Air Force Base, Fla., February 4–6, 2009.

Helmus, Todd C., Christopher Paul, and Russell W. Glenn, *Enlisting Madison Avenue: The Marketing Approach to Earning Popular Support in Theaters of Operation*, Santa Monica, Calif.: RAND Corporation, MG-607-JFCOM, 2007. As of August 3, 2013:
http://www.rand.org/pubs/monographs/MG607.html

Henry, Hugh, "A Non-Kinetic Effects Federate for Training Simulations," *Journal of Defense Modeling and Simulation: Applications, Methodology, Technology*, Vol. 6, No. 3, 2009.

Hitch, Charles J., *Decision-Making for Defense*, Berkeley, Calif.: University of California Press, 1965.

Hodermarsky, George T., and Brian Kalamaja, Science Applications International Corporation, "A Systems Approach to Assessments: Dealing with Complexity," briefing, 2008.

Holdsworth, David R., "Agent-Based Simulation in an Integrated Analysis Process: An Example," presentation at Improving Cooperation Among Nations in Irregular Warfare Analysis, Naval Postgraduate School, Monterey, Calif., December 10, 2007.

Hossack, Andrew, "Strategic Success Factors in Counter-Insurgency Campaigns," presentation at Improving Cooperation Among Nations in Irregular Warfare Analysis, Naval Postgraduate School, Monterey, Calif., December 10, 2007.

Howard, Trevor, "Operational Analysis Support to OP HERRICK," briefing, Defence Science and Technology Laboratory, UK Ministry of Defence, February 13, 2007.

Human Terrain System, "Human Terrain Team Handbook," Fort Leavenworth, Kan., September 2008.

Hume, Robert, director, International Security Assistance Force Afghan Assessment Group, interview with the author, May 7, 2010.

Hunerwadel, J. P., "The Effects-Based Approach to Operations: Questions and Answers," *Air and Space Power Journal*, Spring 2006.

Hylton, David, "Special Inspector General Report on Afghanistan," NATO Training Mission-Afghanistan weblog, June 29, 2010. As of June 30, 2011:
http://ntm-a.com/wordpress2/?p=1732

Iklé, Fred Charles, *Every War Must End*, revised ed., New York: Columbia University Press, 2005.

Intelligence, Surveillance, and Reconnaissance Task Force, "Open Sharing Environment: Unity Net," briefing, July 2010a.

———, "ISR TF Data Sharing Lessons," briefing, presented at Allied Information Sharing Strategy Support to International Security Assistance Force Population Metrics and Data Conference, Brunssum, Netherlands, September 1, 2010b.

Interagency Counterinsurgency Initiative, *U.S. Government Counterinsurgency Guide*, Washington, D.C.: Bureau of Political-Military Affairs, U.S. Department of State, January 2009. As of June 30, 2011:
http://www.state.gov/documents/organization/119629.pdf

Interagency Language Roundtable, homepage, undated. As of June 30, 2011:
http://www.govtilr.org/

International Security Assistance Force, "ISAF Commander Tours U.S. Detention Facility in Parwan," press release, undated. As of June 30, 2011:
http://www.isaf.nato.int/article/isaf-releases/isaf-commander-tours-u.s.-detention-facility-in-parwan.html

International Security Assistance Force Afghan Assessment Group, "Transfer of Lead Security Responsibility Effect Scoring Model," briefing, undated.

International Security Assistance Force Headquarters, *COMISAF's Initial Assessment (Unclassified)*, August 30, 2009.

———, "Knowledge Management," briefing, August 2010.

International Security Assistance Force Headquarters Strategic Advisory Group, "Unclassified Metrics," Kabul, Afghanistan, April 2009.

International Security Assistance Force Joint Command, "District Assessment Framework Tool," spreadsheet, 2010.

International Security Assistance Force Joint Command Chief of Staff, Operational Orders Assessment Steering Group, briefing, November 19, 2009.

International Security Assistance Force Knowledge Management Office, "Afghan Mission Network," briefing, September 1, 2010.

Introduction to the Pacification Data Bank, U.S. government pamphlet, November 1969.

Irregular Warfare Methods, Models and Analysis Working Group, "Final Report," scripted brief, Ft. Leavenworth, Kan.: TRADOC Analysis Center, August 18, 2008.

Irwin, Lewis G., "Irregular Warfare Lessons Learned: Reforming the Afghan National Police," *Joint Force Quarterly*, No. 52, 1st Quarter, 2009.

Jackson, Jack, and MAJ Jon Alt, "Irregular Warfare (IrW) Research to Represent Civilian Populations in Stability Operations," presentation at Military Operations Research Society Conference: Irregular Warfare II, MacDill Air Force Base, Fla., February 4–6, 2009.

Jacobson, Alvin L., and N. M. Lalu, "An Empirical and Algebraic Analysis of Alternative Techniques for Measuring Unobserved Variables," in Hubert M. Blalock, Jr., ed., *Measurement in the Social Sciences: Theories and Strategies*, Chicago, Ill.: Aldine Publishing Company, 1974, pp. 215–242.

Jaiswal, N. K., *Military Operations Research: Quantitative Decision Making*, Boston, Mass.: Kluwer Academic Publishers, 1997.

Johnson, Neil F., *Simple Complexity: A Clear Guide to Complexity Theory*, Oxford, UK: OneWorld, 2009.

Johnson, Thomas H., "The Taliban Insurgency and an Analysis of Shabnamah (Night Letters)," *Small Wars and Insurgencies*, Vol. 18, No. 3, September 2007.

Joint U.S. Public Affairs Office, *Long An Province Survey, 1966*, Saigon, South Vietnam, 1967.

Joint Warfighting Center, U.S. Joint Forces Command, J9, Standing Joint Force Headquarters, *Commander's Handbook for an Effects-Based Approach to Joint Operations*, Suffolk, Va., February 24, 2006a.

———, "Irregular Warfare Special Study," U.S. Joint Forces Command, August 4, 2006b.

Jones, Chad, "Operationalizing Key Leader Engagement: Adapting the Targeting Cycle to Win Friends and Influence People," *Fires*, September-October 2010.

Jones, MAJ Douglas D., U.S. Army, *Understanding Measures of Effectiveness in Counterinsurgency Operations*, Fort Leavenworth, Kan.: School of Advanced Military Studies, U.S. Army Command and General Staff College, May 2006. As of June 30, 2011:
http://handle.dtic.mil/100.2/ADA450589

Kagan, Frederick, W., *Finding the Target: The Transformation of American Military Policy*, New York: Encounter Books, 2006.

Kahler, Mary K., "Providing S-2 Support for a Brigade Support Battalion," *Army Logistician: Professional Bulletin of United States Army Logistics*, November-December 2008.

Kalyvas, Stathis N., and Matthew Adam Kocher, "The Dynamics of Violence in Vietnam: An Analysis of the Hamlet Evaluation System (HES)," *Journal of Peace Research*, Vol. 46, No. 3, May 2009.

Kane, Tim, "Global U.S. Troop Deployment, 1950–2005," Heritage Foundation, Center for Data Analysis, May 24, 2006. As of June 30, 2011:
http://www.heritage.org/research/reports/2006/05/global-us-troop-deployment-1950-2005

Kaplan, Edward H., Moshe Kress, and Roberto Szechtman, "Confronting Entrenched Insurgents," *Operations Research*, Vol. 58, No. 2, March-April 2010.

Kaplan, Lawrence S., Ronald D. Landa, and Edward J. Drea, *The McNamara Ascendancy: 1961– 1965*, Washington, D.C.: Historical Office of the Office of the Secretary of Defense, 2006.

Karnow, Stanley, *Vietnam: A History*, New York: Penguin Books, 1984.

Keaney, Thomas A., "Surveying Gulf War Airpower," *Joint Force Quarterly*, Autumn 1993.

Keefe, Ryan, and Thomas Sullivan, *Resource-Constrained Spatial Hot Spot Identification*, Santa Monica, Calif.: RAND Corporation, TR-768-RC, 2011. As of August 3, 2013:
http://www.rand.org/pubs/technical_reports/TR768.html

Kelly, Justin, and David Kilcullen, "Chaos Versus Predictability: A Critique of Effects-Based Operations," *Security Challenges*, Vol. 2, No. 1, April 2006. As of June 30, 2011: http://www.securitychallenges.org.au/ArticlePages/vol2no1KellyandKilcullen.html

Kem, Jack D., "Understanding the Operational Environment: The Expansion of DIME," *Military Intelligence Professional Bulletin*, Vol. 33, No. 2, April-June 2007.

Kibbey, Benjamin, "1st Infantry Division Recognizes Benefits of Logistics Reporting Tool," *Army Sustainment*, November-December 2010.

Kiesling, Eugenia C., "On War Without the Fog," *Military Review*, Vol. 81, No. 5, September-October 2001.

Kilcullen, David, *The Accidental Guerrilla: Fighting Small Wars in the Midst of a Big One*, Oxford, UK: Oxford University Press, 2009a.

———, "Measuring Progress in Afghanistan," Kabul, Afghanistan, December 2009b.

———, *Counterinsurgency*, New York: Oxford University Press, 2010.

Kinnard, Douglas, *The War Managers*, Hanover, N.H.: University of New Hampshire Press, 1977.

Kinner, Scott, "Expanding Attack the Network," *Air Land Sea Bulletin*, September 2012. As of May 30, 2013: http://www.alsa.mil/library/alsb/ALSB%202012-3.pdf

Kipp, Jacob, "The Human Terrain System: A CORDS for the 21st Century," *Military Review*, Vol. 86, No. 5, September-October 2006.

Kober, Avi, "The Israeli Defense Forces in the Second Lebanon War: Why the Poor Performance?" *Journal of Strategic Studies*, Vol. 31, No. 1, February 2008, pp. 3–40.

Komer, R. W., news conference, December 1, 1967. As of January 27, 2011: http://www.virtual.vietnam.ttu.edu

———, "Text of Ambassador Komer's News Conference," transcript, January 24, 1968a. As of April 1, 2011: http://www.vietnam.ttu.edu

———, "Memorandum for the Record," official memorandum, November 5, 1968b (BACM Research, Disk 2).

———, *Impact of Pacification on Insurgency in South Vietnam*, Santa Monica, Calif.: RAND Corporation, P-4443, August 1970. As of June 30, 2011: http://www.rand.org/pubs/papers/P4443.html

Kruglanski, Arie W., and Donna M. Webster, "Motivated Closing of the Mind: 'Seizing' and 'Freezing,'" *Psychological Review*, Vol. 103, No. 2, 1996.

Kugler, Cornelius W., *Operational Assessment in a Counterinsurgency*, Newport, R.I.: Naval War College, May 10, 2006.

Lamb, Christopher, *Review of Psychological Operations: Lessons Learned from Recent Operational Experience,* Washington, D.C.: National Defense University, September 2005.

Langer, Gary, briefing on ABC News poll of Afghan perceptions, presented at the Center for Strategic and International Studies, December 2009.

Larimer, Larry, John Checco, and Jay Persons, "Irregular Warfare Methods, Models and Analysis Working Group: Final Report," TRADOC Analysis Center, TRAC-F-TR-08-035, August 18, 2008.

Larson, Eric, Richard E. Darilek, Daniel Gibran, Brian Nichiporuk, Amy Richardson, Lowell H. Schwartz, and Cathryn Quantic Thurston, *Foundations of Effective Influence Operations: A Framework for Enhancing Army Capabilities*, Santa Monica, Calif.: RAND Corporation, MG-654-A, 2009a. As of August 3, 2013:
http://www.rand.org/pubs/monographs/MG654.html

Larson, Eric V., Richard E. Darilek, Dalia Dassa Kaye, Forrest E. Morgan, Brian Nichiporuk, Diana Dunham-Scott, Cathryn Quantic Thurston, and Kristin J. Leuschner, *Understanding Commanders' Information Needs for Influence Operations*, Santa Monica, Calif.: RAND Corporation, MG-656-A, 2009b. As of August 3, 2013:
http://www.rand.org/pubs/monographs/MG656.html

Larson, Eric V., Derek Eaton, Brian Nichiporuk, and Thomas S. Szayna, *Assessing Irregular Warfare: A Framework for Intelligence Analysis*, Santa Monica, Calif., RAND Corporation, MG-668-A, 2008. As of August 3, 2013:
http://www.rand.org/pubs/monographs/MG668.html

Law, Averill M., and W. David Kelton, *Simulation Modeling and Analysis*, 1st ed., McGraw-Hill Higher Education, 1982.

Lewy, Guenter, *America in Vietnam*, New York: Oxford University Press, 1978.

Liddell Hart, B. H., *Strategy*, New York: Praeger, 1967.

Lidy, A. Martin, "Evolving USG Interagency Intervention Doctrine—Requirements for Supporting Methods and Tools," presentation at MORS Workshop: Analysis for Non-Traditional Security Challenges: Methods and Tools, JHU/APL, Laurel, Md., February 21–23, 2006.

Lindow, David, Kerry Lenninger, and Brianne Adams, "Methods, Models, and Simulations Research for the Analysis of Stability Operations," presentation at MORS Conference: Irregular Warfare II, MacDill Air Force Base, Fla., February 5, 2009.

Lofdahl, Corey, *Implementing Irregular Warfare Policy Using Modeling and Simulation*, (AIT), BAE Systems Advanced Information Technology, 2009.

Luck, Gary, and Mike Findlay, *Joint Operations Insights and Best Practices*, 2nd ed., Joint Warfighting Analysis Center, U.S. Joint Forces Command, July 2008.

Lukens, Mark W., "Strategic Analysis of Irregular Warfare," Strategy Research Project at the U.S. Army War College, January 3, 2010.

Mahnken, Thomas G., *Technology and the American Way of War Since 1945*, New York: Colombia University Press, 2008.

Mann, Edward C. III, Gary Endersby, and Thomas R. Searle, *Thinking Effects: Effects-Based Methodology for Joint Operations*, Maxwell Air Force Base, Ala.: Air University Press, CADRE paper 15, October 2002.

Mann, Morgan G., "Thoughts Regarding the Company-Level Intelligence Cell," *Marine Corps Gazette*, June 2009.

Mansoor, Peter R., "Linking Doctrine to Action: A New COIN Center-of-Gravity Analysis," *Military Review*, September-October 2007.

Mao Tse-Tung, *On Guerrilla Warfare*, Samuel B. Griffith, trans., Urbana, Ill.: University of Illinois Press, 2000.

Marlowe, Ann, "Defeating IEDs with Data," blog post, *World Affairs Journal*, March 11, 2010. As of June 30, 2011:
http://www.worldaffairsjournal.org/blog/ann-marlowe/defeating-ieds-data

Marr, Jack, "Human Terrain Mapping: A Critical First Step to Winning the COIN Fight," *Military Review*, Vol. 88, No. 2, March-April 2008.

Marshall, S. L. A., *Vietnam Primer: Lessons Learned*, Washington, D.C.: Headquarters, U.S. Army, 1966. As of June 30, 2011:
http://www.lzcenter.com/Documents/12891741-Army-Vietnam-Primer-Pamphlet.pdf

Marvin, Brett L., working group presentation at the U.S. Naval War College, Providence, R.I., October 19, 2010.

Matthews, Lloyd J., and Dale E. Brown, eds., *Assessing the Vietnam War: A Collection from the Journal of the U.S. Army War College*, McLean, Va.: Pergamon-Brassey's International Defense Publishers, 1987.

Matthews, Matt M., *We Were Caught Unprepared: The 2006 Hezbollah-Israeli War*, Fort Leavenworth, Kan.: U.S. Army Combined Arms Center, Combat Studies Institute Press, 2008. As of June 30, 2011:
http://purl.access.gpo.gov/GPO/LPS104476

Mattis, James N., U.S. Marine Corps, "Assessment of Effects Based Operations," memorandum for U.S. Joint Forces Command, Norfolk, Va., August 14, 2008. As of June 30, 2011:
http://smallwarsjournal.com/documents/usjfcomebomemo.pdf

McChrystal, Stanley A., *COMISAF's Initial Assessment (Unclassified),* August 30, 2009.

McCrabb, Maris, "Effects-Based Operations: An Overview," briefing for the Air Force Research Laboratory, January 10, 2008.

McFadden, Willie J. II, and Daniel J. McCarthy, "Policy Analysis," in Andrew G. Loerch and Larry B. Rainey, eds., *Methods for Conducting Military Operational Analysis*, Alexandria, Va.: Military Operations Research Society, 1998.

McFate, Montgomery, "Anthropology and Counterinsurgency: The Strange Story of Their Curious Relationship," *Military Review*, Vol. 85, No. 2, March-April 2005.

McGee, Michael Calvin, "The 'Ideograph': A Link Between Rhetoric and Ideology," *Quarterly Journal of Speech*, Vol. 66, No. 1, February 1980.

McKnight, Patrick E., Katherine M. McKnight, Souraya Sidani, and Aurelio Jose Figueredo, *Missing Data: A Gentle Introduction*, New York: Guilford Press, 2007.

McLear, Michael, *The Ten Thousand Day War, Vietnam: 1945–1975*, New York: St. Martin's Press, 1981.

McLeroy, Carrie, "History of Military Gaming," *Soldiers Magazine*, August 27, 2008. As of May 22, 2012:
http://www.army.mil/article/11936/History_of_Military_gaming/

McMaster, H. R., *Dereliction of Duty: Lyndon Johnson, Robert McNamara, the Joint Chiefs of Staff, and the Lies That Led to Vietnam*, New York: HarperCollins, 1997.

———, "Crack in the Foundation: Defense Transformation and the Underlying Assumption of Dominant Knowledge in Future War," student issue paper, U.S. Army War College, November 2003. As of June 30, 2011:
http://www.au.af.mil/au/awc/awcgate/army-usawc/mcmaster_foundation.pdf

McNamara, Robert S., transcript of a news conference, July 11, 1967. As of January 27, 2011:
http://www.vietnam.ttu.edu

————, "Southeast Asia Operations: Statement by Secretary of Defense McNamara Before Senate Armed Services Committee (excerpts)," testimony on the fiscal year 1969–1973 defense program and the 1969 defense budget, February 1, 1968. As of April 1, 2011:
http://www.vietnam.ttu.edu

————, "Affidavit of Robert S. McNamara," *General William C. Westmoreland v. CBS Inc., et al.*, December 1, 1983. As of March 30, 2011:
http://www.vietnam.ttu.edu

McNamara, Robert S., with Brian VanDeMark, *In Retrospect: The Tragedy and Lessons of Vietnam*, New York: Times Books, 1995.

McNamara, Robert S., James G. Blight, and Robert K. Brigham, with Thomas J. Biersteker and Herbert Y. Schandler, *Argument Without End: In Search of Answers to the Vietnam Tragedy*, New York: Public Affairs, 1999.

Meeks, Heber S., and Barton T. Brundige, "The Role of Intelligence in Sustainment Operations," *Army Sustainment*, Vol. 42, No. 1, January-February 2010.

Meharg, Sarah Jane, *Measuring Effectiveness in Complex Operations: What Is Good Enough*, Calgary, Alberta: Canadian Defence and Foreign Affairs Institute, October 2009. As of June 30, 2011:
http://www.cdfai.org/PDF/Measuring%20Effectiveness%20in%20Complex%20Operations.pdf

Menkhaus, Kenneth J., "State Fragility as a Wicked Problem," *Prism*, Vol. 1, No. 2, March 2010, pp. 85–100.

Metz, Thomas F., Mark W. Garrett, James E. Hutton, Timothy W. Bush, "Massing Effects in the Information Domain: A Case Study in Aggressive Information Operations," *Military Review*, May-June 2006.

Miller, David, and Tom Mills, "Counterinsurgency and Terror Expertise: The Integration of Social Scientists into the War Effort," *Cambridge Review of International Affairs*, Vol. 23, No. 2, 2010.

Miller, Delbert C., and Neil J. Salkind, *Handbook of Research Design and Social Measurement*, 6th ed., Thousand Oaks, Calif.: Sage Publications, 2002.

Miller, John H., and Scott E. Page, *Complex Adaptive Systems: An Introduction to Computational Models of Social Life*, Princeton, N.J.: Princeton University Press, 2007.

Ministry of Supply and War Office: Military Operational Research Unit, successors and related bodies, reports and papers WO 291, undated.

Mission Coordinator, U.S. Embassy, Saigon, South Vietnam, *MACCORDS Field Reporting System*, July 1, 1969. As of January 27, 2011:
http://www.vietnam.ttu.edu

Mitchell, Robbyn, "Mattis Takes Over as CENTCOM Chief," *St. Petersburg Times*, August 12, 2010. As of June 30, 2011:
http://www.tampabay.com/news/mattis-takes-over-as-centcom-chief/1114800

Mlakar, Joe, "Applying Crime Mapping and Analysis Techniques to Forecast Insurgent Attacks in Iraq," presentation at Military Operations Research Society Conference: Irregular Warfare II, MacDill Air Force Base, Fla., February 4–6, 2009.

Modeling and Simulation Coordination Office, *Modeling and Simulation Glossary,* web page, undated. As of August 12, 2012:
http://www.msco.mil/MSGlossary.html

Monday, Paul, "Architecture of the Counter Insurgency Experiment," *Journal of Defense Modeling and Simulation: Applications, Methodology, Technology*, Vol. 6, No. 2, 2009.

Morgan, Wesley, "Afghanistan Order of Battle," Washington, D.C.: Institute for the Study of War, June 2012.

Morse, Phillip M., and George E. Kimball, *Methods for Military Operations Research*, Cambridge Mass.: M.I.T. Press, 1951.

Motley, Larry L., "Developing a Fuel Management Information System in Iraq," *Army Sustainment*, September-October 2011.

Mowery, Samuel, Warfighting Analysis Division, J8, U.S. Department of Defense, "A System Dynamics Model of the FM 3-24 COIN Manual," briefing, 2009. As of June 30, 2011: http://www.guardian.co.uk/news/datablog/2010/apr/29/mcchrystal-afghanistan-powerpoint-slide

Moyar, Mark, *A Question of Command: Counterinsurgency from the Civil War to Iraq*, New Haven, Conn.: Yale University Press, 2009.

Multinational Force–Iraq, "Charts to Accompany the Testimony of General David H. Petraeus," briefing slides, September 10–11, 2007. As of April 1, 2011: http://smallwarsjournal.com/documents/petraeusslides.pdf

———, "Charts to Accompany the Testimony of General David H. Petraeus," briefing slides, April 8–9, 2008. As of April 1, 2011: http://www.defense.gov/pdf/Testimony_Handout_Packet.pdf

Munoz, Arturo, Assessing Military Information Operations in Afghanistan, Santa Monica, Calif.: RAND Corporation, RB-9659-MCIA, 2012. As of August 3, 2013: http://www.rand.org/pubs/research_briefs/RB9659.html

Murray, William S., "A Will to Measure," *Parameters*, Autumn 2001. As of June 30, 2011: http://www.carlisle.army.mil/USAWC/Parameters/Articles/01autumn/Murray.htm

Nagl, John A., *Counterinsurgency Lessons from Malaya and Vietnam: Learning to Eat Soup with a Knife*, Westport, Conn.: Praeger Publishers, 2002.

Nakashima, Ellen, and Craig Whitlock, "With Air Force's Gorgon Drone, 'We Can See Everything,'" *Washington Post*, January 2, 2011. As of January 15, 2011: http://www.washingtonpost.com/wp-dyn/content/article/2011/01/01/AR2011010102690.html

National Commission for the Protection of Human Subjects of Biomedical and Behavioral Research, *The Belmont Report: Ethical Principles and Guidelines for the Protection of Human Subjects Research*, Washington, D.C.: U.S. Government Printing Office, 1978. As of March 10, 2011: http://www.videocast.nih.gov/pdf/ohrp_belmont_report.pdf

National Intelligence Council, *Estimative Products on Vietnam: 1948–1975*, Pittsburgh, Pa.: U.S. Government Printing Office, April 2005.

Nitschke, Stephen G., *Vietnam: A Complex Adaptive Perspective*, U.S. Marine Corps Command and Staff College, thesis, 1997.

North Atlantic Treaty Organisation, "Decision Support to the Combined Joint Task Force and Component Commanders," Report prepared by the Research and Technology Organisation, Analysis and Simulation Panel, TR-SAS-044, December 2004.

Oak Ridge National Laboratories, LandScan web page, undated. As of August 12, 2013: http://web.ornl.gov/sci/landscan/

Oberdorfer, Don, *Tet! The Turning Point in the Vietnam War*, Baltimore, Md.: Johns Hopkins University Press, 2001.

Office of the Director of National Intelligence, *Analytic Standards*, Intelligence Community Directive No. 203, June 21, 2007. As of June 30, 2011:
http://www.dni.gov/files/documents/ICD/ICD%20203%20Analytic%20Standards%20pdf-unclassified.pdf

Office of the Secretary of Defense, "Metrics Conference Outbrief to IJC/IDC," briefing, March 18, 2010.

Office of the Special Inspector General for Afghanistan Reconstruction, Actions Needed to Improve the Reliability of Afghan Security Force Assessments, SIGAR Audit 10-11, Arlington, Va., June 29, 2010. As of June 30, 2011:
http://www.washingtonpost.com/wp-srv/hp/ssi/wpc/sigar.pdf?sid=ST2010062805531

———, Progress Made Toward Increased Stability Under USAID's Afghanistan Initiative-East Program but Transition to Long Term Development Effort Not Yet Achieved, SIGAR Audit-12-11 Contractor Performance and Oversight/Stabilization Initiative, June 29, 2012.

Office of the Special Inspector General for Iraq Reconstruction, Challenges in Obtaining Reliable and Useful Data on Iraqi Security Forces Continue, SIGIR-09-002, October 21, 2008.

———, Iraqi Security Forces: Police Training Program Developed Sizeable Force, but Capabilities are Unknown, SIGIR 11-003, October 25, 2010.

Olson, James S., and Randy Roberts, *Where the Domino Fell: America and Vietnam, 1945–1990*, New York: St. Martin's Press, 1991.

Oswalt, Ivar, Tim Cooley, William Waite, Elliot Waite, Steve "Flash" Gordon, Richard Severinghaus, Jerry Feinberg, and Gary Lightner, "Calculating Return on Investment for U.S. Department of Defense Modeling and Simulation," *M&S Journal*, Fall 2012.

Otero, Christopher, "Reflections on Clausewitz and Jomini: A Discussion on Theory, MDMP, and Design in the Post OIF Army," *Small Wars & Insurgencies*, May 25, 2011.

Ottenberg, Michael, "Assessing Irregular Warfare: GWOT X Game," presentation at Wargaming and Analysis Workshop, Northrop Grumman Heritage Conference Center, Chantilly, Va., October 16–18, 2007.

Owens, Bill, with Ed Offley, *Lifting the Fog of War*, Baltimore, Md.: Johns Hopkins University Press, 2000.

Paganini, John, "Counterinsurgency Lessons Learned," U.S. Department of Defense, November 16, 2011.

Palmer, Bruce, Jr., *The 25-Year War: America's Military Role in Vietnam*, Lexington, Ky.: University Press of Kentucky, 1984.

Parkman, Jon, and Hanley, Nathan, *Peace Support Operations Model Functional Specifications (PSOM-FS)*, Dstl/TR28869/1.0a, August 13, 2008.

Paschall, Joseph F., "IO for JOE: Applying Strategic IO at the Tactical Level," *FA Journal*, July-August 2005.

Pattee, Phillip G., "Force Protection Lessons from Iraq," *Joint Force Quarterly*, No. 37, 2005.

Paul, Christopher, "Framing the Problem: Challenges to Successful Psychological Operations," *MORS Conference: Irregular Warfare II*, MacDill Air Force Base, Fla., February 4, 2009.

Paul, Christopher, Colin Clarke, and Beth Grill, *Victory Has a Thousand Fathers: Sources of Success in Counterinsurgency*, Santa Monica, Calif.: RAND Corporation, MG-964-OSD, 2010. As of June 30, 2011:
http://www.rand.org/pubs/monographs/MG964.html

Paul, Christopher, Daniel F. McCaffrey, Sarah Fox, "A Cautionary Case Study of Approaches to the Treatment of Missing Data," *Statistical Methods and Applications*, Springer-Verlag, January 8, 2008.

Peck, Michael, "Firmer Ground," *Training and Simulation*, Vol. 14, October 1, 2011.

Perez, Celestino, "A Practical Guide to Design: A Way to Think About It, and a Way to Do It," *Military Review*, Vol. 91, No. 2, March-April 2011.

Perry, Walter L., "Linking Systems Performance and Operational Effectiveness," in Andrew G. Loerch and Larry B. Rainey, eds., *Methods for Conducting Military Operational Analysis*, Alexandria, Va.: Military Operations Research Society, 2007.

———, email exchange with the author, February 3, 2011.

Perry, Walter L., Robert W. Button, Jerome Bracken, Thomas Sullivan, and Jonathan Mitchell, *Measures of Effectiveness for the Information-Age Navy: The Effects of Network-Centric Operations on Combat Outcomes*, Santa Monica, Calif.: RAND Corporation, MR-1449-NAVY, 2002. As of June 30, 2011:
http://www.rand.org/pubs/monograph_reports/MR1449.html

Perry, Walter L., and John Gordon IV, *Analytic Support to Intelligence in Counterinsurgencies*, Santa Monica, Calif.: RAND Corporation, MG-682-OSD, 2008. As of June 30, 2011:
http://www.rand.org/pubs/monographs/MG682.html

Petraeus, David H., "Learning Counterinsurgency: Observations from Soldiering in Iraq," *Military Review*, Vol. 86, No. 1, January-February 2006.

Petty, Mikel D., et al., "A Re-Use Lexicon: Terms, Units, and Modes in M&S Asset Re-use," *M&S Journal*, Fall 2011.

Phillips, Rufus, *Why Vietnam Matters: An Eyewitness Account of Lessons Not Learned*, Annapolis, Md.: Naval Institute Press, 2008.

Piedmont, John P., "Det One, U.S. Marine Corps U.S. Special Operations Command Detachment, 2003–2006: U.S. Marines in the Global War on Terrorism," History Division, United States Marine Corps, 2010.

Pool, Ithiel de Sola, Gordon Fowler, Peter McGrath, and Richard Peterson, *Hamlet Evaluation System Study*, Cambridge, Mass.: Simulmatics Corporation, May 1, 1968. As of June 30, 2011:
http://handle.dtic.mil/100.2/AD839821

Prince, W. B., and J. H. Adkins, *Analysis of Vietnamization: Hamlet Evaluation System Revisions*, Ann Arbor, Mich.: Bendix Aerospace Systems Division, February 1973. As of June 30, 2011:
http://www.dtic.mil/cgi-bin/GetTRDoc?AD=AD908686&Location=U2&doc=GetTRDoc.pdf

Proposal Review: Predictive Analysis Tools Assessment, Actionable Hot Spot Monitoring (AHSM), Building Time-Sensitive Clusters in Time and Space, PATA Technical Review, undated.

Public Law 110-28, U.S. Troop Readiness, Veterans' Care, Katrina Recovery, and Iraq Accountability Appropriations Act, May 25, 2007.

Pusateri, Anthony E., "Metrics to Monitor Governance and Reconstruction in Afghanistan: Development of Measures of Effectiveness for Civil-Military Operations and a Standardized Tool to Monitor Governance Quality," United States Army Civil Affairs and Psychological Operations Command, Technical Report 04-01, March 12, 2004.

Quade, E. S., ed., *Analysis for Military Decisions*, Santa Monica, Calif.: RAND Corporation, R-387-PR, 1964. As of June 30, 2011:
http://www.rand.org/pubs/reports/R0387.html

Quinlivan, James T., "Force Requirements in Stability Operations" *Parameters*, Winter 1995.

Race, Jeffrey, *War Comes to Long An: Revolutionary Conflict in a Vietnam Province*, Berkeley and Los Angeles, Calif.: University of California Press, 1972.

Ramjeet, T. J., "Operational Analysis in Support of HQ RC(S), Kandahar, Afghanistan, September 2007 to January 2008," in *The Cornwallis Group XIII, Analysis in Support of Policy*, Farnborough, UK: Defence Science and Technology Laboratory, 2008.

Record, Jeffrey, *Hollow Victory: A Contrary View of the Gulf War*, New York: Brassey's, Inc., 1993.

Rehm, Allan, ed., "Analyzing Guerrilla Warfare," conference proceedings, September 24, 1985. As of January 15 2011:
http://www.virtual.vietnam.ttu.edu

Rennie, Ruth, Sudhindra Sharma, and Pawan Kumar Sen, *Afghanistan in 2009: A Survey of the Afghan People*, Kabul, Afghanistan: Asia Foundation, December 2009. As of June 30, 2011:
http://www.asiafoundation.org/resources/pdfs/Afghanistanin2009.pdf

"Rescinding ASCOPEs for PMESII-PT at the Tactical Level; Possibly Good in Theory, but What About in Application?" *U.S. Army Combined Arms Center*, web blog, April 20, 2009. As of January 25, 2010:
http://usacac.army.mil/blog/blogs/coin/archive/2009/04/20/rescinding-ascopes-for-pmesii-pt-at-the-tactical-level-possibly-good-in-theory-but-what-about-in-application.aspx

Ricchiardi, Sherry, "Whatever Happened to Iraq?" *American Journalism Review*, June–July 2008. As of November 9, 2010:
http://www.ajr.org/article.asp?id=4515

Richardson, Damon B., Richard F. Deckro, and Victor D. Wiley, "Modeling and Analysis of Post-Conflict Reconstruction," *The Journal of Defense Modeling and Simulation: Applications, Methodology, Technology*, Vol. 1, No. 4, 2004.

Rittel, Horst W. J., and Melvin M. Webber, "Dilemmas in a General Theory of Planning," *Policy Sciences*, Vol. 4, 1973, pp. 155–169.

Rogers, Simon, "The McChrystal Afghanistan PowerPoint Slide: Can You Do Any Better?" *Guardian Data Blog*, April 29, 2010. As of June 30, 2011:
http://www.guardian.co.uk/news/datablog/2010/apr/29/mcchrystal-afghanistan-powerpoint-slide

Rosenau, William, and Austin Long, *The Phoenix Program and Contemporary Counterinsurgency*, Santa Monica, Calif.: RAND Corporation, OP-258-OSD, 2009. As of June 30, 2011:
http://www.rand.org/pubs/occasional_papers/OP258.html

Roush, Maurice D., "The Hamlet Evaluation System," *Military Review*, September 1969.

Roy-Bhattacharya, Joydeep, *The Storyteller of Marrakesh*, New York: W. W. Norton and Company, 2011.

Rubin, Alissa J., "U.S. Report on Afghan War Finds Few Gains in 6 Months," *New York Times*, April 29, 2010a. As of June 30, 2011:
http://www.nytimes.com/2010/04/30/world/asia/30afghan.html

———, "Petraus Says Taliban Have Reached Out to Karzai," *New York Times*, September 27, 2010b. As of June 30, 2011:
http://www.nytimes.com/2010/09/28/world/asia/28afghan.html

Ruby, Tomislav Z., "Effects-Based Operations: More Important Than Ever," *Parameters*, Autumn 2008, pp. 26–35.

Rumsfeld, Donald H., interview with Steve Inskeep, *Morning Edition*, National Public Radio, March 29, 2005. As of June 30, 2011:
http://www.defense.gov/transcripts/transcript.aspx?transcriptid=2551

Russell, James A., "Innovation in War: Counterinsurgency Operations in Anbar and Ninewa Provinces, Iraq, 2005–2007," *Journal of Strategic Studies*, Vol. 33, No. 4, 2012.

Saaty, Thomas L., *Mathematical Methods of Operations Research*, New York: McGraw-Hill, 1959.

———, "A Model of 21st Century Counterinsurgency Warfare," *The Journal of Defense Modeling and Simulation: Applications, Methodology, Technology*, Vol. 4, No. 3, 2007.

Schott, LTC Russel, "Irregular Warfare: Building a Counterinsurgency based Tactical-Level Analytic Capability," presentation at Military Operations Research Society Conferences Conference: Irregular Warfare II, MacDill Air Force Base, Fla., February 4–6, 2009.

Schrecker, Mark, "U.S. Strategy in Afghanistan: Flawed Assumptions Will Lead to Ultimate Failure," *Joint Force Quarterly*, No. 59, 4th Quarter, 2010, pp. 75–82.

Schroden, Jonathan J., "Measures for Security in a Counterinsurgency," *Journal of Strategic Studies*, Vol. 32, No. 5, October 2009, pp. 715–744.

———, email exchange with the author, February 1, 2010a.

———, email exchange with the author, September 29, 2010b.

———, briefing at the U.S. Naval War College, Newport, R.I., October 19, 2010c.

———, email exchange with the author, October 27, 2010d.

———, "Why Operations Assessments Fail: It's Not Just the Metrics," *Naval War College Review*, Autumn 2011.

Seagrist, Thomas, "Combat Advising in Iraq: Getting Your Advice Accepted," *Military Review*, May-June 2010.

Shadid, Anthony, "In Iraq, Western Clocks, but Middle Eastern Time," *New York Times*, August 14, 2010. As of June 30, 2011:
http://www.nytimes.com/2010/08/15/weekinreview/15shadid.html

Sharp, U. S. G., and William C. Westmoreland, *Report on the War in Vietnam as of 30 June 1968*, Washington, D.C.: U.S. Government Printing Office, 1968.

Shea, Dennis P., and Julianne B. Nelson, "Towards a Business Model to Encourage Re-use of Models and Simulations in DoD," *M&S Journal*, Fall 2011.

Shearer, Robert, "Operations Analysis in Iraq: Sifting Through the Fog of War," *Military Operations Research*, Vol. 16, No. 2, 2011.

Sheehan, Neil, *A Bright Shining Lie: John Paul Vann and America in Vietnam*, New York: Vintage Books, 1988.

Shepherd, John, "ERIS: Exploratory Regional Insurgency Simulation," presentation at MORS Workshop: Analysis for Non-Traditional Security Challenges: Methods and Tools, JHU/APL, Laurel, Md., February 21–23, 2006.

Shields, S. Scott, "The Case Study Approach: The Use of Case Studies in Information Operations," presentation at Military Operations Research Society Conference: Irregular Warfare II, MacDill Air Force Base, Fla., February 4–6, 2009.

Shimko, Keith L., *The Iraq Wars and America's Military Revolution*, New York: Cambridge University Press, 2010.

Shmorrow, Dylan, "Sociocultural Behavior Research and Engineering in the Department of Defense Context," Office of the Secretary of Defense, Assistant Secretary of Defense for Research and Engineering Human Performance, Training, and BioSystems Directorate, August 2011.

Shrader, Charles R., *History of Operations Research in the United States Army,* Vol. 1: *1942–62,* Washington, D.C.: Office of the Deputy Under Secretary of the Army for Operations Research, U.S. Army, Center for Military History publication 70-102-1, August 11, 2006. As of June 30, 2011: http://purl.access.gpo.gov/GPO/FDLP521

Siegl, Michael B., "Sustaining a BCT in Southern Iraq," *Army Sustainment,* November-December 2010.

Smith, David, "CAA Current Operations Support to OIF/OEF," Presented at the 49th Army Operations Research Symposium, October 13–14 2010. (Document obtained from MCCDC Operations Analysis Current Operations Analysis Support Team [COAST] on May 11, 2012.)

Smith, Douglas S., commander, 2nd Battalion, 47th Infantry, 9th Infantry Division, interview with Robert L. Keeley, commander, 19th Military History Detachment, Bien Phuoc, Republic of Vietnam, July 1, 1969. As of June 30, 2011: http://www.history.army.mil/documents/vietnam/vnit/vnit457f.htm

Smith, Edward Allen, *Effects-Based Operations: Applying Network Centric Warfare to Peace, Crisis, and War,* Washington, D.C.: U.S. Department of Defense Command and Control Research Program, November 2002.

Smith, Neil, "Retaking Sa'ad: Successful," *Armor,* July-August 2007.

Soeters, Joseph, "Measuring the Immeasurable? The Effects-Based Approach in Comprehensive Peace Operations," draft revision of a paper presented at the tenth European Research Group on Military and Society conference, Stockholm, Sweden, June 23, 2009.

Sok, Sam, "ORSA: Operations Research Systems Analysts Help Develop Solutions," *Infantry,* September-October, 2011.

Sok, Sang Am, Center for Army Analysis, "Assessment Doctrine," briefing presented at the Allied Information Sharing Strategy Support to ISAF Population Metrics and Data Conference, Brunssum, Netherlands, September 1, 2010.

Sokolowski, John A., Catherine M. Banks, and Brent Morrow, "Using an Agent-Based Model to Explore Troops Surge Strategy," *Journal of Defense Modeling and Simulation: Applications, Methodology, Technology,* Vol. 1, No. 14, 2011.

Sollinger, Jerry D., U.S. Army psychological operations advisor to Phong Dinh Province, Vietnam, in 1967, interview with the author, Washington, D.C., April 19, 2010.

Sorley, Lewis, ed., *The Abrams Tapes: 1968–1972,* Lubbock, Tex.: Texas Tech University Press, 2004.

St. Romain III, Plauche J., "Tactical-Level Intelligence During Counterinsurgency Campaigns," *The Officer,* Vol. 86, No. 1, February-March 2010.

State of Washington, "Recidivism of Adult Felons 2004," web page, December 2005. As of May 10, 2011: http://www.sgc.wa.gov/PUBS/Recidivism/Adult_Recidivism_Cy04.pdf

Stolzenberg, Ross M., ed., *Sociological Methodology,* Vol. 26, 2006.

Strong, Paul, "The Peace Support Operations Model: Strategic Interaction Process," *Journal of Defense Modeling and Simulation: Applications, Methodology, Technology,* Vol. 8, No. 2, 2011.

Sullivan, Thomas J., Iraq Reconstruction Management Office, U.S. Embassy Baghdad, "Long Term Plan for IRMO Metrics," memorandum, August 2004.

Sweeney, LTC Brian, "An IO Perspective on Afghanistan: An Appreciation for Tribal Complexity," presentation at Military Operations Research Society Conference: Irregular Warfare II, MacDill Air Force Base, Fla., February 4, 2009.

Sweetland, Anders, *Item Analysis of the HES (Hamlet Evaluation System)*, Santa Monica, Calif.: RAND Corporation, D-17634-ARPA/AGILE, 1968. As of June 30, 2011: http://www.rand.org/pubs/documents/D17634.html

Sykes, Charles S., *Interim Report of Operations, First Cavalry Division: July 1965 to December 1966*, Albuquerque, N.M.: First Cavalry Division Association, undated. As of January 15, 2011: http://www.virtual.vietnam.ttu.edu

Talbot, Oliver, and Noel Wilde, "Modeling Security Sector Reform Activities in the Context of Stabilization Operations," *The Journal of Defense Modeling and Simulation: Applications, Methodology, Technology*, Vol. 8, No. 2, 2011.

Terriff, Terry, Frans Osinga, and Theo Farrell, *A Transformation Gap? American Innovations and European Military Change*, Stanford, Calif.: Stanford University Press, 2010.

Thayer, Thomas C., ed., *A Systems Analysis View of the Vietnam War, 1965–1972*, Vol. 1: *The Situation in Southeast Asia*, Washington, D.C.: Office of the Assistant Secretary of Defense for Systems Analysis, Southeast Asia Intelligence Division, 1975a.

———, *A Systems Analysis View of the Vietnam War, 1965–1972*, Vol. 9: *Population Security*, Washington, D.C.: Office of the Assistant Secretary of Defense for Systems Analysis, Southeast Asia Intelligence Division, 1975b.

———, *A Systems Analysis View of the Vietnam War, 1965–1972*, Vol. 10: *Pacification and Civil Affairs*, Washington, D.C.: Office of the Assistant Secretary of Defense for Systems Analysis, Southeast Asia Intelligence Division, 1975c.

———, *War Without Fronts: The American Experience in Vietnam*, Boulder, Colo.: Westview Press, 1985.

Thiel, Joshua, "Debunking the 10 to 1 Ratio and Surges," *Small Wars Journal*, January 15, 2011.

Tho, Tran Dinh, *Pacification*, Indochina Monographs, Washington, D.C.: U.S. Army Center of Military History, 1980. As of June 30, 2011: http://cgsc.cdmhost.com/cdm/fullbrowser/collection/p4013coll11/id/1409/rv/singleitem/rec/2

Thompson, Robert, *No Exit from Vietnam*, New York: David McKay Company, Inc., 1970.

Toevank, Freek-Jan, "Methods for Collecting and Sharing Data," plenary presentation, Allied Information Sharing Strategy Support to ISAF Population Metrics and Data Conference, Brunssum, Netherlands, September 1, 2010.

Tolk, Andreas, et al., "Towards Methdological Approaches to Meet the Challenges of Human, Social, Cultural, and Behavioral (HSCB) Modeling," Proceedings of the 2010 Winter Simulation Conference, IEEE, 2010.

TRADOC—*See* U.S. Army Training and Doctrine Command.

Trinquier, Roger, *Modern Warfare: A French View of Counterinsurgency*, trans. Daniel Lee, New York: Frederick A. Praeger, 1964.

Tufte, Edward R., *The Visual Display of Quantitative Information*, 2nd ed., Cheshire, Conn.: Graphics Press, 2006.

Tunney, John V., *Measuring Hamlet Security in Vietnam: Report of a Special Study Mission*, Washington, D.C.: U.S. Government Printing Office, December 1968.

Ulrich, Mark, U.S. Army and Marine Corps Joint Counterinsurgency Center, "Center of Influence Analysis: Linking Theory to Application," briefing presented at the National Defense University, 2009.

U.S. Agency for International Development, Office of Military Affairs (OMA), "Tactical Conflict Assessment and Planning Framework (TCAPF)," briefing, undated.

————, *Measuring Fragility: Indicators and Methods for Rating State Performance*, Washington, D.C., June 2005. As of June 30, 2011:
http://pdf.usaid.gov/pdf_docs/PNADD462.pdf

————, "Stabilization and the District Stability Framework," briefing slides, USAID Office of Military Affairs, July 7, 2010.

U.S. Army, "FM 5–0 Overview," Combined Arms Directorate, briefing, April 1, 2010.

U.S. Army Combat Studies Institute, *Wanat: Combat Action in Afghanistan, 2008*, Fort Leavenworth, Kan.: Combat Studies Institute Press, US Army Combined Arms Center, 2008.

U.S. Army Combined Arms Center, "Understanding the Operational Environment in COIN," briefing, February 13, 2009.

U.S. Army Training and Doctrine Command, "Commander's Appreciation and Campaign Design," Version 1.0, Pamphlet 525-5-500, January 28, 2008. As of June 29, 2011:
http://www.tradoc.army.mil/tpubs/pams/p525-5-500.pdf

————, G-2 Intelligence Support Activity, "Operational Environment Laboratory Overview," briefing, September 2012. As of October 31, 2012:
http://www.onesaf.net/community/systemdocuments/UsersConference2010/1_Tuesday/1110%20
-%20TRISA%20-%20Operational%20Environment%20Lab%20Overview%20-%20UC2010%20
-%20Jordan.pdf

U.S. Army Training and Doctrine Command Analysis Center, Senior Analyst Review of IWTWG, September 15, 2011.

————, "TRAC Irregular Warfare Tactical Wargame Update," Briefing for the IEA 1448 Meeting, March 14, 2012.

U.S. Army War College, *How The Army Runs: A Senior Leader Reference Handbook, 2011–2012*, Carlisle, Penn., 2012.

U.S. Bureau of Labor Statistics, "How the Government Measures Unemployment," web page, October 16, 2009. As of June 30, 2011:
http://www.bls.gov/cps/cps_htgm.htm

U.S. Central Command, "CJOA—Afghanistan TF 236 Overview," unpublished briefing, September 1, 2010.

U.S. Central Intelligence Agency, *Analysis of the Vietnamese Communists' Strengths, Capabilities, and Will to Persist in Their Present Strategy in Vietnam, Annex VI: The Morale of the Communist Forces*, August 26, 1966. As of March 15, 2011:
http://www.vietnam.ttu.edu/star/images/041/04114192001e.pdf

————, *The Communist's Ability to Recoup Their Tet Military Losses*, memorandum, March 1, 1968. As of March 15, 2011:
http://www.vietnam.ttu.edu/star/images/041/0410586008.pdf

————, *A Compendium of Analytic Tradecraft Notes*, Vol. 1, Washington, D.C., February 1997. As of June 30, 2011:
http://www.au.af.mil/au/awc/awcgate/cia/tradecraft_notes/contents.htm

————, A Tradecraft Primer: Structured Analytic Techniques for Improving Intelligence Analysis, March 2009. As of June 30, 2011:
https://www.cia.gov/library/publications/publications-rss-updates/tradecraft-primer-may-4-2009.html

U.S. Defense Logistics Agency, *A Study of Strategic Lessons Learned in Vietnam,* Vol. VI: *Conduct of the War, Book 2: Functional Analysis,* May 2, 1980. As of June 30, 2011:
http://handle.dtic.mil/100.2/ADA096430

U.S. Department of Defense, *DoD Modeling and Simulation (M&S) Glossary*, DOD 5000.59-M, Under Secretary of Defense for Acquisition Technology, January 1998.

————, *Measuring Security and Stability in Iraq*, Washington, D.C., November 2006. As of January 13, 2011:
http://www.defense.gov/home/features/iraq_reports/index.html

————, *Measuring Security and Stability in Iraq*, Washington, D.C., December 2007. As of June 30, 2011:
http://www.defense.gov/home/features/iraq_reports/index.html

————, *Report on Progress Toward Security and Stability in Afghanistan: Report to Congress in Accordance with the 2008 National Defense Authorization Act (Section 1230, Public Law 110-181)*, Washington, D.C., semiannual report 2, January 2009a.

————, *Measuring Stability and Security in Iraq*, Washington, D.C., September 2009b. As of June 30, 2011:
http://www.defense.gov/home/features/iraq_reports

————, "COMISAF Initial Assessment (Unclassified)," Republished in the Washington Post, September 21, 2009c. As of August 8, 2012:
http://www.washingtonpost.com/wp-dyn/content/article/2009/09/21/AR2009092100110.html

————, *Report on Progress Toward Security and Stability in Afghanistan: Report to Congress in Accordance with the 2008 National Defense Authorization Act (Section 1230, Public Law 110-181)*, Washington, D.C., April 2010a.

————, *Report on Progress Toward Security and Stability in Afghanistan: Report to Congress in Accordance with the 2008 National Defense Authorization Act (Section 1230, Public Law 110-181)*, Washington, D.C., November 2010b.

————, *Measuring Stability and Security in Iraq*, Washington, D.C., June 2010c. As of June 30, 2011:
http://www.defense.gov/home/features/iraq_reports

————, "DoD News Briefing with Vice Adm. Harward from Afghanistan," November 30, 2010d. As of May 10, 2011:
http://www.globalsecurity.org/military/library/news/2010/11/mil-101130-dod02.htm

————, *Report on Progress Toward Security and Stability in Afghanistan: Report to Congress in Accordance with the 2008 National Defense Authorization Act (Section 1230, Public Law 110-181)*, Washington, D.C., October 2011.

————, *Sustaining U.S. Global Leadership: Priorities for 21st Century Defense*, January 2012. As of July 19, 2012:
http://www.defense.gov/news/Defense_Strategic_Guidance.pdf

U.S. Department of Defense Directive, *DoD Modeling and Simulation (M&S) Management*, Number 5000.59, Under Secretary of Defense for Acquisition, Technology, and Logistics, August 8, 2007.

————, "Support for Strategic Analysis (SSA)," Number 8260.05, July 7, 2011.

U.S. Department of Defense Office of Force Transformation, *Military Transformation: A Strategic Approach*, Washington, D.C., 2003.

———, *The Implementation of Network-Centric Warfare*, Washington, D.C., January 5, 2005. As of September 16, 2010:
http://purl.access.gpo.gov/GPO/LPS57633

U.S. Department of Defense Office of the Inspector General's Inspection and Evaluation Directorate, "Combined Forces Command—Afghanistan Management Decision Model," Arlington, Va., 2005.

U.S. District Court, Southern District of New York, *General William C. Westmoreland v. CBS Inc., et al.*, March 3, 1983.

U.S. Government Interagency Counterinsurgency Initiative, *U.S. Government Counterinsurgency Guide*, Washington, D.C.: U.S. Department of State, Bureau of Political-Military Affairs, January 2009. As of June 30, 2011:
http://www.state.gov/documents/organization/119629.pdf

U.S. Joint Chiefs of Staff, *Joint Logistics*, Joint Publication 4-0, Washington, D.C., July 18, 2006a. As of November 15, 2012:
http://www.dtic.mil/doctrine/new_pubs/jp4_0.pdf

———, *Joint Operation Planning*, Joint Publication 5–0, Washington, D.C., December 26, 2006b. As of June 27, 2011:
http://www.dtic.mil/doctrine/new_pubs/jp5_0.pdf

———, *Joint Intelligence*, Joint Publication 2-0, Washington, D.C., June 22, 2007. As of June 27, 2011:
http://www.dtic.mil/doctrine/new_pubs/jp2_0.pdf

———, Joint Doctrine and Education Division Staff, "Effects-Based Thinking in Joint Doctrine," *Joint Force Quarterly*, No. 53, 2nd Quarter 2009a, p. 60.

———, *Joint Intelligence Preparation of the Operational Environment*, Joint Publication 2-01.3, Washington, D.C., June 16, 2009b.

———, *Counterinsurgency Operations*, Joint Publication 3-24, Washington, D.C., October 5, 2009c. As of June 27, 2011:
http://www.dtic.mil/doctrine/new_pubs/jp3_24.pdf

———, *Joint Operations*, Joint Publication 3-0, incorporating change 2, Washington, D.C., March 22, 2010. As of June 27, 2011:
http://www.dtic.mil/doctrine/new_pubs/jp3_0.pdf

———, *Department of Defense Dictionary of Military and Associated Terms*, Joint Publication 1-02, Washington, D.C., as amended through May 15, 2011. As of June 27, 2011:
http://purl.access.gpo.gov/GPO/LPS14106

U.S. Marine Corps, *Military Occupational Specialties Marine Corps Manual*, Marine Corps Order 1200.17, May 23, 2008.

U.S. Marine Corps Intelligence Activity, *Cultural Generic Information Requirements Handbook (C-GIRH)*, August 2008.

———, *Cultural Intelligence Indicators Guide*, Marine Corps Intelligence Activity, October 2009.

U.S. Military Assistance Command, Vietnam, *CORDS Report—Phong Dinh Province 2/68*, province monthly report, February 1968a. As of June 30, 2011:
http://www.vietnam.ttu.edu/star/images/212/2121008002.pdf

———, *MACCORDS-OAD Fact Sheet: RD Cadre Evaluation System*, November 15, 1968b.

———, *Commander's Summary of the MACV Objectives Plan*, 1969a.

———, *Directive on MAC-CORDS Field Reporting System*, July 1, 1969b.

———, *System for Evaluating the Effectiveness of Republic of Vietnam Armed Forces (SEER)*, November 27, 1969c. As of January 15, 2011:
http://arcweb.archives.gov

———, *Form Used for Hamlet Evaluation Survey (HES), Report by CORDS Employees*, 1969d.

———, *Hamlet Evaluation System District Advisors' Handbook*, 1971a. As of January 27, 2011:
http://www.vietnam.ttu.edu

———, *The Hamlet Evaluation System (HES) Reports*, official memorandum, January 22, 1971b.

———, *Hamlet Evaluation System Advisors Handbook*, June 1971c.

———, *Order of Battle Summary*, Vols. 1 and 2, March 1972.

U.S. National Institute of Justice, "Measuring Recidivism," web page, February 20, 2008. As of May 10, 2011:
http://www.nij.gov/topics/corrections/recidivism/measuring.htm

U.S. Naval Postgraduate School, *NPS Academic Catalog*, September 10, 2010. As of June 30, 2011:
http://www.nps.edu/admissions/Doc/Academic_Catalog_10_SEPT_2010.pdf

U.S. Senate Foreign Relations Committee, *Vietnam Hearings: Answers to Questions for the Hearing Record on the CORDS Program*, transcript, March 1970.

USAID—*See* U.S. Agency for International Development.

Usmani, Zeeshan-ul-hassan, and Daniel Kirk, "Modeling and Simulation of Explosion Effectiveness as a Function of Blast and Crowd Characteristics," *Journal of Defense Modeling and Simulation: Applications, Methodology, Technology*, Vol. 6, No. 2, 2009.

Vanden Brook, Tom, "IEDs Kill 21,000 Iraqi Civilians 2005–2010," *USA Today*, January 11, 2011. As of January 13, 2011:
http://www.usatoday.com/news/world/iraq/2011-01-12-1Aied12_ST_N.htm

Van Riper, Paul K., "EBO: There Was No Baby in the Bathwater," *Joint Force Quarterly*, No. 52, 1st Quarter 2009.

Vego, Milan N., "Effects-Based Operations: A Critique," *Joint Force Quarterly*, No. 41, 2nd Quarter 2006.

Vincent, Etienne, Philip Eles, and Boris Vasiliev, "Opinion Polling in Support of Counterinsurgency," in *The Cornwallis Group XIV: Analysis of Societal Conflict and Counter-Insurgency*, Cornwallis Group, 2010. As of April 1, 2011:
http://www.thecornwallisgroup.org/cornwallis_2009/7-Vincent_etal-CXIV.pdf

Vines, John R., "The XVIII Airborne Corps on the Ground in Iraq," *Military Review*, September-October 2008.

von Bertalanffy, Ludwig, *General System Theory: Foundations, Development, Applications*, revised ed., New York: G. Braziller, 1974.

von Clausewitz, Carl, *On War*, J. J. Graham, trans., London: N. Trübner, 1874. As of June 30, 2011:
http://www.netlibrary.com/urlapi.asp?action=summary&v=1&bookid=2009343

Wagenhals, Lee W., Alex Levis, and Sajjad Haider, *Planning, Execution, and Assessment of Effects-Based Operations (EBO)*, Rome, N.Y.: Air Force Research Laboratory, AFRL-IF-RS-TR-2006-176, May 2006. As of June 30, 2011:
http://handle.dtic.mil/100.2/ADA451493

Walker, David M., *DoD Should Provide Congress and the American Public with Monthly Data on Enemy-Initiated Attacks in Iraq in a Timely Manner*, Washington, D.C.: U.S. Government Accountability Office, GAO-07-1048R, September 28, 2007. As of June 30, 2011:
http://purl.access.gpo.gov/GPO/LPS87929

Warren, Gemma, and Patrick Rose, "Representing Strategic Communication and Influence in Stabilization Modeling," *The Journal of Defense Modeling and Simulation: Applications, Methodology, Technology*, Vol. 8, No. 2, 2011.

Wass de Czege, Huba, "Systemic Operational Design: Learning and Adapting in Complex Missions," *Military Review*, January-February 2009. As of June 30, 2011:
http://usacac.army.mil/CAC2/MilitaryReview/Archives/English/MilitaryReview_20090228_art004.pdf

Watts, Stephen, "Political Dilemmas of Stabilization and Reconstruction," in Paul K. Davis, ed., *Dilemmas of Intervention: Social Science for Stabilization and Reconstruction*, Santa Monica, Calif.: RAND Corporation, MG-1119-OSD, 2012. As of August 3, 2013:
http://www.rand.org/pubs/monographs/MG1119.html

Wendt, Eric P, "Strategic Counterinsurgency Modeling," *Special Warfare*, Vol. 18, No. 2, September 2005.

Westmoreland, William C., "General Westmoreland's Activities Report for September," memorandum for the President of the United States, October 10, 1967a.

―――, "Military Briefing by General William Westmoreland, USA, Commander, Military Assistance Command, Vietnam," November 22, 1967b.

Wheeler, Earle G., Chairman, U.S. Joint Chiefs of Staff, "Khe Sanh," memorandum to the President of the United States, February 3, 1968. As of June 27, 2011:
http://www.vietnam.ttu.edu/star/images/001/0010113001A.pdf

White House, *Iraq Benchmark Assessment Report*, Washington, D.C.: U.S. Government Printing Office, September 17, 2007.

Wilder, Andrew, "Losing Hearts and Minds in Afghanistan," *Viewpoints—Afghanistan, 1979–2009: In the Grip of Conflict*, Washington, D.C.: Middle East Institute, December 2, 2009. As of June 30, 2011:
http://www.mei.edu/content/losing-hearts-and-minds-afghanistan

Wilkes, Stuart, "Counterinsurgency Wargame Development: Estimating Security Force Requirements," presentation at Wargaming and Analysis Workshop, Northrop Grumman Heritage Conference Center, Chantilly, Va., October 16–18, 2007.

Williams-Bridgers, Jacquelyn, *Iraq and Afghanistan: Security, Economic, and Governance Challenges to Rebuilding Efforts Should be Addressed in U.S. Strategies*, testimony before the U.S. House of Representatives Committee on Armed Services, March 25, 2009, Washington, D.C.: U.S. Government Accountability Office, GAO-09-476T, 2009. As of June 30, 2011:
http://purl.access.gpo.gov/GPO/LPS113358

"Working Group 1: Data and Knowledge Management," presentation at Improving Analytical Support to the Warfighter: Campaign Assessments, Operational Analysis, and Data Management, Lockheed Martin Missiles & Fire Control, Orlando, Fla., April 19–22, 2010.

"Working Group 2: Campaign Assessments," presentation at Improving Analytical Support to the Warfighter: Campaign Assessments, Operational Analysis, and Data Management, Lockheed Martin Missiles & Fire Control, Orlando, Fla., April 19–22, 2010.

"Working Group 3: Operational and Tactical Assessments," presentation at Improving Analytical Support to the Warfighter: Campaign Assessments, Operational Analysis, and Data Management, Lockheed Martin Missiles & Fire Control, Orlando, Fla., April 19–22, 2010.

"Working Group 4: Current Ops Analysis—Tactical," presentation at Improving Analytical Support to the Warfighter: Campaign Assessments, Operational Analysis, and Data Management, Lockheed Martin Missiles & Fire Control, Orlando, Fla., April 19–22, 2010.

"Working Group 5: Current Operations Analysis—Strategic & Operational Level," presentation at Improving Analytical Support to the Warfighter: Campaign Assessments, Operational Analysis, and Data Management, Lockheed Martin Missiles & Fire Control, Orlando, Fla., April 19–22, 2010.

Wray, John D., *Optimizing Helicopter Assault Support in a High Demand Environment*, Thesis, Naval Postgraduate School, Monterey, Calif., June 2009.

Zacharias, Greg L., Jean MacMillan, and Susan B. Van Hemel, *Behavior Modeling and Simulation: From Individuals to Societies*, National Research Council, 2008.

Zehna, Peter W., ed., *Selected Methods and Models in Military Operations Research*, rev. ed., Honolulu, Hawaii: University Press of the Pacific, 2005.

Zhai, Qiang, *China and the Vietnam Wars, 1950–1975*, Chapel Hill, N.C.: University of North Carolina Press, 2000.

Zhu, Hongwei, and Richard Y. Yang, "An Information Quality Framework for Verifiable Intelligence Products," Massachusetts Institute of Technology, undated. As of June 30, 2011: http://mitiq.mit.edu/Documents/Publications/Papers/2007/An%20IQ%20Framework%20for%20 Verifiable%20Intelligence%20Products.pdf

Ziliak, Stephen Thomas, and Dierdre N. McCloskey, *The Cult of Statistical Significance: How the Standard Error Costs Us Jobs, Justice, and Lives*, Ann Arbor, Mich.: University of Michigan Press, 2008.